Lecture Notes in Mathematics

Edited by A. Dold and B. Eckmann

656

Probability Theory on Vector Spaces

Proceedings, Trzebieszowice, Poland, September 1977

Edited by A. Weron

Springer-Verlag
Berlin Heidelberg New York 1978

Editor
A. Weron
Institute of Mathematics
Wroclaw Technical University
Wybrzeże Wyspianskiego 27
50-370 Wrocław
Poland

AMS Subject Classifications (1970): 28A40, 28A45, 47A20, 47A45, 47D05, 60B05, 60B15, 60E05, 60F05, 60F20, 60G10, 60G15, 60G17, 60G25, 60G45

ISBN 3-540-08846-6 Springer-Verlag Berlin Heidelberg New York
ISBN 0-387-08846-6 Springer-Verlag New York Heidelberg Berlin

Printed in Germany

Printing and binding: Beltz Offsetdruck, Hemsbach/Bergstr.
2141/3140-543210

F o r e w o r d

In this volume we present the extended versions of most lectures given at the Conference on Probability Theory on Vector Spaces, held at Trzebieszowice in the mountains region of Lower Silesia, Poland, September 1 - 7, 1977. The conference was organized by the Institute of Mathematics of Wroclaw Technical University and was attended by over 80 registered participants from 10 countries.

The conference emphasized the functional analysis aspects of the probability theory. The main topics were: probability measures on some linear spaces /Gaussian and stable measures, martingales and central limit theorem/, infinite-dimensional stochastic processes and their connection with the dilation theory.

The organizers wish to thank Professor S.Gładysz, the director of the Institute of Mathematics for his lively interest and help in the preparation of the conference. We also express our gratitude for all the people who by their help contributed to the good scientific and friendly atmosphere of the conference. Special thanks are due to Mrs. O.Szeląg, Mr. J.Górniak and Mr. P.Kajetanowicz for their help in organizing the conference.

Let us gratefully acknowledge the oral and written contributions of all the participants, who helped make the conference a succes. Last but not least we want to express our appreciation to Mrs. A.Peters, maths. editor of Springer-Verlag for her kind assistance during the whole period of preparation of this volume, and to Mrs. W.Cupiał for typing all manuscripts.

Aleksander Weron

C o n t e n t s

Lectures not included in this volume

Weak convergence of measures on $Lip_{\alpha}[0,\infty]$
Z.ANTOSZEWSKI /Łódź/

Quasi diffusion solutions of stochastic differential equations
J.DĄBROWSKI and A.KOSYK /Warsaw/

A zero one law for Gaussian measures
T.INGLOT /Wrocław/

Operator stable probability measures on Banach spaces
W.KRAKOWIAK /Wrocław/

A characterization of operator semi-stable measures on R^N
A.ŁUCZAK /Łódź/

An isomorphic theorem for Banach space valued stationary sequences
A.MAKAGON /Wrocław/

The domain of attraction of certain stable measures on Banach spaces
M.MARCUS /Evanston/

Cylindrical martingales and stochastic integrals
M.METIVIER /Palaiseau/

On a property of Gaussian measures on Banach spaces
B.MINCER /Wrocław/

An example of the decomposability semigroup
T.RAJBA /Wrocław/

Random integral in Banach spaces and the central limit theorem
J.ROSIŃSKI /Wrocław/

Pre-supports and kernels of probability measures in Frechet spaces
W.SMOLEŃSKI /Warsaw/

An Invariance principle for the law of iterated logarithm for a sequence of random variables
D.SZYNAL /Lublin/

On Banach spaces with the Sazonov property
V.I.TARIELADZE /Tbilisi/

Banach space valued second order processes and the dilation theory
A.WERON /Wrocław/

Pseudomoments
V.M.ZOLOTAREV /Moscow/

SOME RESULTS CONCERNING GAUSSIAN MEASURES ON METRIC LINEAR SPACES

T. Byczkowski

This paper contains results of two types: the zero-one law for Gaussian measures on measurable vector spaces (or even measurable groups) and an investigation of Gaussian measures on more concrete vector spaces, namely on separable Orlicz spaces.

1. Zero-one law for Gaussian measures.

The results of this section are based on a method sketched in the proof of Theorem 2.2 in [5] and generalize that theorem in several directions. We obtain here the zero-one law for completion measurable "rational" subgroups of measurable groups and we indicate that this result can be extended to measurable subgroups of some Polish groups. Results of this section generalize the earlier ones, obtained by Kallianpur [15], Jain [14], Baker [1] and Rajput [17] and contain also the zero-one law for Brownian motion process with values in a separable LCA group.

Definition 1. Let G be an abelian group and let $\mathcal{B}$ be a σ-field of subsets of G. $(G,\mathcal{B})$ is called a measurable group (m.g.) if the addition

$$(x,y) \longrightarrow x + y$$

is measurable with respect to the σ-field $\mathcal{B}$ and the product σ-field $\mathcal{B}\times\mathcal{B}$ and if the inverse mapping

$$x \longrightarrow -x$$

is measurable with respect to $\mathcal{B}$.

The typical example of a m.g. is a metric separable group with its Borel σ-field. Also the space $D[0,1]$ of all left-continuous real functions defined on the unit interval, without discontinuities of the second kind, with the Borel σ-field (with respect to the Skorochod topology) is a m.g., although it is not a topological group (see [2]). Arbitrary vector space G with the σ-field generated by a vector space of linear functionals yields another example of a m.g. .

Roughly speaking, a m.g. is a group on which the convolution $\mu * \vartheta$ can be defined for arbitrary pair of probability measures μ, ϑ on $(G, \mathcal{B})$:

$$\mu * \vartheta(A) = \mu \times \vartheta(\{(x,y) \; ; \; x + y \in A\})$$

for every $A \in \mathcal{B}$, where $\mu \times \vartheta$ denotes the product of μ and ϑ.

A mapping X, defined on a probability space (Ω, Σ, P), with values in a m.g. $(G, \mathcal{B})$ will be called a random element (r.e.) if it is measurable with respect to Σ and $\mathcal{B}$.

If X and Y are two r.e.'s defined on a common probability space (Ω, Σ, P) with values in a m.g. then $X + Y$, $X - Y$ are also r.e.'s. If X, Y are independent in the usual sense, then the distribution of $X + Y$ equals $\mu * \vartheta$, where μ, ϑ are distributions of X, Y, respectively.

If $(G, \mathcal{B})$ is a m.g. then we can consider Gaussian measures on G.

<u>Definition 2.</u> Let $(G, \mathcal{B})$ be a m.g.. A probability measure μ on $(G, \mathcal{B})$ is called Gaussian if for every pair of independent r.e.'s X_1, X_2 having the distribution μ the r.e.'s

$$X_1 + X_2 \quad \text{and} \quad X_1 - X_2$$

are independent.

Let $\Psi : G \times G \longrightarrow G \times G$ be defined by the following formula:

$$\Psi(x,y) = (x+y, \; x-y).$$

Then the above definition can be stated equivalently (Ψ is clearly $\mathcal{B} \times \mathcal{B}$ measurable) :

μ in Gaussian iff there are probability measures ϑ_1, ϑ_2 such that

$$\Psi(\mu \times \mu) = \vartheta_1 \times \vartheta_2 \; ,$$

(that is such that

$$\mu \times \mu(\Psi^{-1}(A)) = \vartheta_1 \times \vartheta_2(A)$$

for every $A \in \mathcal{B} \times \mathcal{B}$).

The measures ϑ_1, ϑ_2 are, in fact, uniquley determined by μ :

$$\vartheta_1(A) = \vartheta_1 \times \vartheta_2(A \times G) = \mu \times \mu(\{(x,y); \; x+y \in A\}) =$$
$$= \mu * \mu(A)$$

for $A \in \mathcal{B}$, so $\vartheta_1 = \mu * \mu$. Analogously, $\vartheta_2 = \mu * \bar{\mu}$, where $\bar{\mu}(B) = \mu(-B)$ for $B \in \mathcal{B}$.

If $(G, \mathcal{B})$ is a separable real topological vector space with the Borel σ-field, such that G has sufficiently many continuous linear functionals, then this definition is equivalent to the usual one: μ is Gaussian if and only if for every continuous linear functional f, $f(\cdot)$ is a real Gaussian random variable on the probability space $(G, \mathcal{B}, \mu)$. Definition 2 has been used by Fréchet in[13] in case Banach spaces and by Corvin in [10] in case LCA groups.

Now, if μ is a probability measure on a measurable space $(G, \mathcal{B})$ then the completion of $\mathcal{B}$ with respect to μ we will denote by $\widetilde{\mathcal{B}}^{\mu}$. A $\mathcal{B}$-measurable mapping $\varphi: G \longrightarrow G$ will be called bi-measurable if $\varphi(A) \in \mathcal{B}$ whenever $A \in \mathcal{B}$.

Now, we are ready to formulate our result.

<u>Theorem 1.</u> Let $(G, \mathcal{B})$ be a m.g. and let μ be a Gaussian measure on $(G, \mathcal{B})$. If F is a $\mathcal{B}$-measurable subgroup of G such that G/F is torsion-free (that is, does not contain any element of finite order) then

$$\mu(F) = 0 \quad \text{or} \quad \mu(F) = 1.$$

If $x \longrightarrow 2x$ is a bi-measurable mapping then the above statement holds for $\widetilde{\mathcal{B}}^{\mu}$-measurable subgroups F of G having the property that G/F is torsion-free.

If $(G, \mathcal{B})$ is a m.g. such that G is a Polish space, $\mathcal{B}$ is its Borel σ-field, and G does not contain elements of order 2 then $x \longrightarrow 2x$ is bi-measurable, in virtue of Kuratowski's Theorem.

In the proof of Theorem 1 we use two lemmas. The arguments used in the proof of the first one are similar to those used in Lemma 1 in [11].

<u>Lemma 1.</u> Let $(G, \mathcal{B})$ be a m.g. such that $x \longrightarrow 2x$ is bi-measurable. Let μ be a Gaussian measure on $(G, \mathcal{B})$ and let F be a $\widetilde{\mathcal{B}}^{\mu}$- measurable subgroup of G, closed under division by 2 (that is $2x \in F \Longrightarrow x \in F$). Assume that $\mu(F) > 0$ and let

$$F' = \{x \in G;\ F + x \in \widetilde{\mathcal{B}}^{\mu}\}$$

Then $F' \in \widetilde{\mathcal{B}}^{\mu}$ and $\mu(F') = 1$.

<u>Proof.</u> Let us denote $\{x \in G;\ 2x \in A\} = 1/2\, A$. Then, by assumption, we have

$$F = 1/2\, F.$$

Since μ is Gaussian, we have

$$\Psi\,(\mu \times \mu) = \vartheta_1 \times \vartheta_2$$

where $\vartheta_1 = \mu * \mu$, $\vartheta_2 = \mu * \bar{\mu}$. Hence

$$\Psi^2(\mu \times \mu) = \Psi(\vartheta_1 \times \vartheta_2).$$

Since $\Psi^2(x,y) = (2x,2y)$, we have

$$\mu(\{x;\ 2x \in A\}) = \mu \times \mu\ (\{(x,y);\ (2x,2y) \in A \times G\})$$

$$= \vartheta_1 \times \vartheta_2(\{(x,y);\ x+y \in A,\ \ x-y \in G\}) =$$

$$\vartheta_1 \times \vartheta_2(\{(x,y);\ x+y \in A\}) = \vartheta_1 * \vartheta_2(A) = \mu^{*3} * \bar{\mu}(A)$$

for every $A \in \mathcal{B}$.

Next, since $F \in \tilde{\mathcal{B}}^{\mu}$ there exist $C_0,\ D_0 \in \mathcal{B}$ such that

$$C_0 \subseteq F \subseteq D_0$$

and $\mu(D_0 \setminus C_0) = 0$. Let

$$C = \bigcup_{n=-\infty}^{\infty} 2^n C_0,\quad D = \bigcap_{n=-\infty}^{\infty} 2^n D_0.$$

Then $C,\ D \in \mathcal{B}$, $1/2\ C = C$, $1/2\ D = D$, $D \setminus C \subseteq D_0 \setminus C_0$ and

$$C \subseteq F \subseteq D.$$

Hence

$$0 = \mu(1/2(D \setminus C)) = \mu(\{x;\ 2x \in D \setminus C\}) = \mu^{*3} * \bar{\mu}(D \setminus C)$$

$$= \int \mu^{*2} * \bar{\mu}((D \setminus C)-x)\ d\mu(x).$$

Let

$$F'' = \{\ x\ ;\ \mu^{*2} * \bar{\mu}((D \setminus C)-x) = 0\}.$$

Then $\mu(F'') = 1$ so

$$\mu(\{x;\ F-x\ \in\ \tilde{\mathcal{B}}^{\mu^{*2} * \bar{\mu}}\}) = 1.$$

Since $\mu(F) > 0$, there is an $x_0 \in F$ such that

$$F-x_0 = F\ \in\ \tilde{\mathcal{B}}^{\mu^{*2} * \bar{\mu}}$$

Hence

$$0 = \mu^{*2} * \bar{\mu}\ (D' \setminus C') = \int \mu * \bar{\mu}((D' \setminus C')-x)\ d\mu(x)$$

for $D',\ C' \in \mathcal{B}$ such that $C' \subseteq F \subseteq D'$, $\mu^{*2} * \bar{\mu}(D' \setminus C') = 0$. By the arguments we have just used we obtain that $F \in \tilde{\mathcal{B}}^{\mu * \bar{\mu}}$. By repeating this argument , we obtain the conclusion.

<u>Lemma 2.</u> Let $(G, \mathcal{B})$ be a countable abelian group with the σ-field $\mathcal{B}$ of all subsets of G. Let μ be a Gaussian measure on $(G, \mathcal{B})$. If $\mu(\{0\}) > 0$ then μ is the normed Haar measure concentrated on some finite subgroup of G.

Proof. If we endow G with the discrete topology then $(G, \mathcal{B})$ is a separable locally compact abelian group with its Borel σ-field $\mathcal{B}$. Let $\hat{G}$ be the character group of G. It is easy to observe that $\hat{\mu}$ (the characteristic function of μ) satisfies the following equation :

$$\hat{\mu}(\chi+\gamma)\ \hat{\mu}(\chi-\gamma) = \hat{\mu}(\chi)^2\ \hat{\mu}(\gamma)\ \hat{\mu}(-\gamma), \text{ for } \chi, \gamma \in \hat{G}.$$

Let φ be the characteristic function of the symmetrization of μ. Then $\varphi(-\gamma) = \varphi(\gamma)$ so we have

$$\varphi(\chi+\gamma)\ \varphi(\chi-\gamma) = \varphi(\chi)^2\ \varphi(\gamma)^2, \quad \text{for } \chi, \gamma \in \hat{G}. \tag{i}$$

Let

$$H = \{\gamma;\ \varphi(\gamma) \neq 0\}.$$

From (i) we obtain that H is a subgroup of $\hat{G}$. Since φ is continuous, H is open (hence closed). We show that

$$H = \{\gamma;\ \varphi(\gamma) = 1\}.$$

Let $\varphi(\gamma) > 0$. From (i) we obtain

$$\varphi(\gamma) = [\varphi(2\gamma)]^{1/4} = \dots = [\varphi(2^n\gamma)]^{1/2^{2n}}.$$

(observe that $\varphi \geq 0$). Since $\hat{G}$ is compact, there is a subsequence n_k and $\gamma_0 \in \hat{G}$ such that $2^{n_k}\gamma \longrightarrow \gamma_0$. Because $2^{n_k}\gamma \in H$ and H is closed, $\gamma_0 \in H$, thus $\varphi(\gamma_0) > 0$. Hence and by the continuity of φ, we obtain $\varphi(\gamma) = 1$. Thus $\varphi(\gamma) \subseteq \{0,1\}$ and φ is the characteristic function of an idempotent measure θ (i.e. such that $\theta * \theta = \theta$). From [16] (ch.III, Th.3.1) it follows that φ is the characteristic function of the normed Haar measure concentrated on a finite subgroup of G. So, we have proved that $\mu * \bar{\mu} = \theta$, where θ is the Haar measure of a finite subgroup I. Now, let $C(\mu)$ denote the support of μ. Let $x_0 \in C(\bar{\mu})$. Then

$$x_0 + C(\mu) \subseteq C(\mu * \bar{\mu}) = I$$

so we have

$$\mu * x_0 * \theta = \theta = \mu * \bar{\mu}.$$

Let $\Delta = \{\gamma;\ \hat{\mu}(\gamma) = 0\}$. Then

$$\Delta = \{\gamma;\ \hat{\bar{\mu}}(\gamma) = 0\} = \{\gamma;\ (\hat{\mu} * \hat{\bar{\mu}})(\gamma) = 0\}$$

hence we infer that

$$\mu * \theta = \mu$$

Thus

$$\mu * x_0 = \mu * x_0 * \theta = \theta.$$

Because of the assumption $\mu(\{0\}) > 0$ we have $0 \in C(\bar{\mu})$ so we can take $x_0 = 0$ and finally we obtain $\mu = \theta$.

Remark. The arguments used in the proof of Lemma 2 are based on those used in [16], p. 101, Remark 2.

Proof of Theorem 1. Let F be a $\tilde{\mathcal{B}}^{\mu}$ measurable subgroup of G. Since G/F is torsion-free, it does not contain, in particular, any non-zero element of order two. Hence F is closed under division by 2. If $F \in \mathcal{B}$, then $F + x \in \mathcal{B}$ for every $x \in G$ since $(G, \mathcal{B})$ is a m.g.. If $F \in \tilde{\mathcal{B}}^{\mu}$ and the mapping $x \longrightarrow 2x$ is bi-measurable, then $F + x \in \tilde{\mathcal{B}}^{\mu}$ for $x \in F'$, where $F' \in \tilde{\mathcal{B}}^{\mu}$ and $\mu(F') = 1$, by Lemma 1. In the former case, observe that $F + F' \supseteq F'$ hence $F + F' \in \tilde{\mathcal{B}}^{\mu}$ and $\mu(F + F') = 1$. Thus, we have two possibilities: $F + x \subseteq F + F'$ and then $F + x \in \tilde{\mathcal{B}}^{\mu}$ or $(F + x) \cap (F + F') = \emptyset$ which implies that $F + x$ is a null set, so $F + x \in \tilde{\mathcal{B}}^{\mu}$. Thus, in both cases, $F + x \in \tilde{\mathcal{B}}^{\mu}$, for every $x \in G$. Now, let $\pi: G \longrightarrow G/F$ be the canonical homomorphism from G onto G/F. Let E_0 be the subgroup of G generated by the set $\{x \,;\, \mu(F+x) > 0\}$. Let

$$E = \bigcup_{n=0}^{\infty} 1/2^n \; E_0 .$$

Then E is a subgroup of G, closed under division by 2. Since there are at most countably many cosets of F, having positive measure μ, E_0 consists of countably many cosets of F and therefore $E_0 \in \tilde{\mathcal{B}}^{\mu}$. Since $\pi(1/2\, E_0) = \{\pi(y);\, 2y \in E_0\} \subseteq \{\pi(y)\,;\, 2\pi(y) \in \pi(E_0)\}$ and $[x] \longrightarrow 2[x]$ is one-to-one, for $[x] \in G/F$, we infer that $\pi(1/2\, E_0)$ is countable. By the induction $\pi(1/2^n\, E_0)$ is countable, hence so is $\pi(E)$. Since $F \subseteq E = E+F$, E also consists of countably many cosets of F, so $E + x \in \tilde{\mathcal{B}}^{\mu}$, for every $x \in G$. Hence, it is clear that $\mu(E+x) = 0$, if $x \notin E$. Using arguments similar to those used in Lemma 1 we obtain that E is $\tilde{\mathcal{B}}^{\vartheta_1 * \vartheta_2}$ measurable and that the following formula holds :

$$\mu(\{x;\, 2x \in E\}) = \vartheta_1 * \vartheta_2(E).$$

Since $\vartheta_1 * \vartheta_2(E) = \int \bar{\mu}(E-x) \; d\mu^{*3}(x) = \int_E \mu(E) \; d\mu^{*3}(x)$

$$= \mu(E)\, \mu^{*3}(E) = \dots = [\mu(E)]^4$$

and $E = 1/2\, E$, we thus obtain

$$0 < \mu(F) \leqslant \mu(E) = \mu(\{x \,;\, 2x \in E\}) = \vartheta_1 * \vartheta_2(E) = \mu(E)^4 .$$

So, we have $\mu(E) = 1$. Next, if we define the measure $\pi(\mu)$ on $\pi(E)$ by the formula :

$$\pi(\mu)\ (\{[x]\}) = \mu(\pi^{-1}([x])) = \mu(F+x), \quad x \in \pi(E)$$

then $\pi(E)$ with the σ-field of all subsets is, of course, a m.g. and $\mu' = \pi(\mu)$ is a Gaussian measure on $(\pi(E), 2^{\pi(E)})$:

$$\mu' \times \mu'(\{([x],[y]);\ (x,y) \in E \times E,\ \ x+y \in u+F,\ x-y \in v+F\}) =$$

$$\mu \times \mu(\{(x,y);\ x+y \in u+F,\ \ x-y \in v+F\}) = \vartheta_1 \times \vartheta_2((u+F)\times(v+F))$$
$$= \vartheta_1'\ (\{[u]\})\ \ \vartheta_2'(\{[v]\}),$$

where $\vartheta_i'\ (\{[u]\}) = \vartheta_i'(\pi^{-1}[u]) = \vartheta_i(F+u)$.

By Lemma 2, μ' is concentrated on a finite subgroup of $\pi(E)$ and, because G/F does not contain any finite, nontrivial subgroup, μ' is degenerated at the identity of $\pi(E)$, that is

$$\mu(F) = 1.$$

Now, let G be a vector space. Let F be a rational subspace of G, which means that $rx+sy \in F$ for each $x,y \in F$ and all rationals r,s. It is clear that G/F is torsion-free. So, we have the following :

<u>Corollary.</u> Let $(G, \mathcal{B})$ be a vector space in which addition and multiplication by $\pm\ 1/2$ are measurable with respect to $\mathcal{B}\times\mathcal{B}$ and $\mathcal{B}$, respectively. Let μ be a Gaussian measure on $(G,\mathcal{B})$ and let F be a $\tilde{\mathcal{B}}^{\mu}$ measurable rational subspace of G. Then $\mu(F) = 0$ or $\mu(F) = 1$.

The following example indicates that, in general, we cannot expect for arbitrary Gaussian measure μ on a m.g. $(G,\mathcal{B})$ that $\mu(F) = 0$ or 1, for every subgroup $F \in \mathcal{B}$.

<u>Example.</u> Let $G = Z(3)$ be the cyclic group of order 3. The normed Haar measure on G is Gaussian in the sense of Definition 1. However, $\mu(\{0\}) = 1/3$.

So, we must exclude the situations indicated by the above example.

<u>Definition 3.</u> A probability measure μ is called without idempotent factors if the equality $\mu * \lambda = \mu$, for an idempotent λ, implies that λ is concentrated at the identity.

Now, we can formulate the next result.

Theorem 2. Let $(G, \mathcal{B})$ be a Polish abelian group with the Borel σ-field $\mathcal{B}$. Assume that every Gaussian measure on G is a translation of a symmetric one. Let μ be a Gaussian measure on G, without idempotent factors. Then for every $\tilde{\mathcal{B}}^{\mu}$ measurable subgroup F of G $\mu(F) = 0$ or $\mu(F) = 1$.

The proof of this theorem will appear in Colloqium Mathematicum. Here, we mention only that the assumption that every Gaussian measure μ on $(G, \mathcal{B})$ is of the form $\vartheta * x_0$, where ϑ is a symmetric measure, is satisfied in many situations : if G is a LCA group satisfying the second countability axiom, with the Borel σ-field $\mathcal{B}$ [10], or if G is a complete separable locally convex vector space with the σ-field generated by all continuous linear functionals on G (then x_0 is, in fact, the expectation of μ - see [8], [12]). This assumption is also satisfied when G is a separable Orlicz space (non-necessarily locally convex) [6]. Finally, it is also satisfied when we consider $C_G[0,1]$ - the group of all continuous functions defined on [0,1] with values in a separable LCA group.

2. Gaussian measures on Orlicz spaces.

In Banach spaces, the usual definition of Gaussian measures is formulated in terms of continuous linear functionals.

Unfortunately, there are metric linear spaces, very natural from the point of view of the probability theory, having no nontrivial continuous linear functionals. The best known are the $L_p[0,1]$ spaces, $0 < p < 1$ (of real functions, defined on the unit interval and p-integrable with respect to the Lebesgue measure), or, more general, some Orlicz spaces $L_\Phi[0,1]$.

So, we have to consider some modification of the notion of the dual space of vector spaces.

This leads to the following definition :

Definition 4. Let $(E, \mathcal{B}, \mu)$ be a measurable real vector space with a probability measure μ. A real $\mathcal{B}$-measurable function f is called an additive measurable functional if

$$f(x \pm y) = f(x) \pm f(y) \quad \mu \times \mu \; - \; \text{a.e.}$$

By Bernstein's Theorem, if μ is Gaussian, then $f(.)$ is a (real) Gaussian random variable.

The problem of the existence of sufficiently many additive measurable functionals is very important and, if the considered vector space G is a function space, this problem is closely related to the problem of the correspondence between measures on G and measurable stochastic processes with sample paths in G.

Now, let $(T,\mathcal{F}, m)$ be a finite measure space and let Φ be a non-decreasing continuous function defined on the right half-line vanishing only at the origin. Let L_0 be the space of all $\mathcal{F}$-measurable real functions on T. For $x \in L_0$ put

$$R_{\Phi}(x) = \int_T \Phi(|x(t)|)\, m(dt)$$

Let L_Φ be the set of all $x \in L_0$ such that $R_\Phi(ax) < \infty$ for a positive constant a. The set L_Φ is a linear space under usual addition and scalar multiplication. Moreover, it becomes a complete metric linear space under (usually non-homogeneous) seminorm $||\cdot||_\Phi$:

$$||x||_\Phi = \inf\{ c;\ \ c > 0,\ \ R_\Phi(c^{-1}x) \leq c\}.$$

The space $(L_\Phi, ||\cdot||_\Phi)$ is called an Orlicz space [18].

In the sequel we will assume that T is a separable metric space, $\mathcal{F}$ its Borel σ-field and that Φ satisfies (Δ_2)condition:

$$\Phi(2u) \leq k\, \Phi(u) \quad \text{for} \quad u \geq u_0$$

for some constants $k > 0$ and $u_0 \geq 0$.

Let $\{\xi(t)\ ;\ t \in T\}$ be a stochastic process defined on a probability space (Ω,Σ,P) ; it is said to be measurable if the mapping $\xi : \Omega \times T \longrightarrow R$ defined by $(\omega,t) \longrightarrow \xi(\omega,t)$ is measurable with respect to $\Sigma \times \mathcal{F}$ and the Borel σ-field $\mathcal{B}_R$ of R. If ξ is measurable then $\xi(\omega,\cdot) \in L_0$, for all $\omega \in \Omega$.

Now, let us suppose that $\xi(\omega,\cdot) \in L_\Phi$ a.e.. Let $\tilde{\xi} : \Omega \longrightarrow L_\Phi$ be defined as follows :

$$\tilde{\xi}(\omega) = \begin{cases} \xi(\omega,\cdot) & \text{if } \xi(\omega,\cdot) \in L_\Phi \\ 0 & \text{if } \xi(\omega,\cdot) \notin L_\Phi \end{cases}$$

Since $\tilde{\xi}$ has a separable range [9], it is a r.e. with values in L_Φ . The probability distribution of $\tilde{\xi}$ will be denoted by μ_ξ and called the measure induced by the process ξ.

The next theorem is a slightly translated version of a result of Chung and Doob (see [9] and Theorem 3 in [19]).

<u>Theorem 3.</u> Let X be a r.e. defined on a probability space (Ω,Σ,P) with values in L_Φ . Then the following are equivalent :

(i) X has a separable range.

(ii) There exists a measurable stochastic process $\xi: \Omega \times T \longrightarrow \bar{R}$ such that $\tilde{\xi} = X$ a.e. .

(iii) There exists a separable (in the sense of Doob) and measurable stochastic process $\xi : \Omega \times T \longrightarrow \bar{R}$ such that $\tilde{\xi} = X$ a.e. .

Proof. In virtue of [9] we need only to show that (ii) implies (i). By the measurability of ξ

$$\xi(\omega,t) = \lim \sum a_i^{(n)} \chi_{H_i^n}(\omega,t) = \lim \xi^{(n)}(\omega, t)$$

$P \times m$-a.e. where $H_i^n \in \Sigma \times \mathcal{F}$. Without any loss of generality we can assume that $|\xi^{(n)}| \leq 2 |\xi|$. By Fubini's Theorem and (Δ_2) condition we infer that $\xi^{(n)}(\omega) \longrightarrow \xi(\omega)$ in L_Φ, P-a.e. . The proof will be completed if we show that every function f of the form

$$f(\omega,t) = \chi_H(\omega,t) \ , \quad H \in \Sigma \times \mathcal{F}$$

induces the mapping $\tilde{f} : \Omega \longrightarrow L_\Phi$ having a separable range. Since every such f is clearly measurable, we infer by [9] that there exists a countable subset of L_0, say $\{x_i\}_{i=1}^{\infty}$ such that $\tilde{f}$ is in its L_0-closure. We can assume that $|x_i(t)| \leq 2$ (for, if not, we can take $y_i = x_i \wedge 2$). But, in this case we have

$$\tilde{f}(\omega) = \lim x_{i_n} \quad \text{in } L_\Phi, \quad \text{P-a.e.}$$

for a certain subsequence of x_i, so $\tilde{f}$ is contained in L_Φ-closure of $\{x_i\}_{i=1}^{\infty}$, which completes the proof.

Now, we can prove that if μ is Gaussian then we can construct a Gaussian measurable stochastic process ξ representing μ.

Theorem 4. Let μ be a Gaussian measure on a separable Orlicz space L_Φ. Then there exists a measurable Gaussian stochastic process ξ on $\Omega \times T$, $(\Omega,\Sigma,P) = (L_\Phi, \mathcal{B}_{L_\Phi}, \mu)$, inducing μ.

Proof. Let $\vartheta = 1/3\ \mu + 1/3\ \mu * \mu + 1/3\ \mu * \bar{\mu}$. Since the space L_Φ is separable, by Theorem 3 there exists a measurable stochastic process η, defined on $L_\Phi \times T$ such that $\tilde{\eta}(x) = x$ ϑ-a.e. . By the definition of ϑ we have

$$\tilde{\eta}(x \pm y) = x \pm y \quad \mu \times \mu\text{-a.e.}$$

and

$$\tilde{\eta}(x) \pm \tilde{\eta}(y) = x \pm y \quad \mu \times \mu\text{-a.e.}$$

Hence

$$\tilde{\eta}(x \pm y) = \tilde{\eta}(x) \pm \tilde{\eta}(y) \quad \mu \times \mu - \text{a.e.}$$

By Fubini's Theorem there exists $T_o \subseteq T$, $T \in \mathcal{F}$, $m(T_o) = 0$ such that if $t \notin T_o$ then

$$\eta(x \pm y,t) = \eta(x,t) \pm \eta(y,t) \quad \mu \times \mu - \text{a.e.} .$$

By Bernstein's Theorem we infer that if $t_1, t_2, t_3, \ldots, t_n \in T \setminus T_o$ then

$$< \xi_{t_1} , \xi_{t_2}, \ldots, \xi_{t_n} >$$

is a Gaussian random vector. Now, if we put $\xi_t = \eta_t$ if $t \notin T_o$ and $\xi_t = 0$ if $t \in T_o$ we obtain a measurable stochastic process ξ having all the required properties.

Observe that the process ξ , appearing in the conclusion of the above theorem can be choosen, in addition, separable (in the sense of Doob), in virtue of Theorem 3.

The proof of this theorem, based on other ideas, has been given by Rajput in [17] (for L_p spaces, $1 \leq p < \infty$). A simple proof of this theorem, without appealing to the Chung - Doob result, is contained in [4] (pp. 193-196).

Now, in order to formulate our next result, we state here some basic facts on Borel structures in Polish spaces. The reader is referred to [16], [3].

Let X be a Polish space and $\mathcal{B}$ its Borel σ-field. A subset of X is called analytic if it is of the form f(Q), where Q is a Polish space and f is a continuous map of Q into X. It is well known that all Borel subsets of X are analytic. Now, let $(X, \mathcal{B})$ be a measurable space. We say that $(X, \mathcal{B})$ is countably generated if there is a countable subfamily of $\mathcal{B}$ which generates $\mathcal{B}$. We say that $\mathcal{B}$ separates points if there is a countable subfamily of $\mathcal{B}$, say $\{E_i\}$, such that if $x,y \in X$, $x \neq y$ then $\chi_{E_i}(x) \neq \chi_{E_i}(y)$ for a positive integer i, χ_E denoting the characteristic function of E. The following theorem, due to Blackwell, will be used in the sequel.

<u>Unique Structure Theorem.</u> Let E be an analytic subset of a Polish space endowed with the relative Borel σ-field $\mathcal{B}$. Let $\mathcal{B}_o$ be a countably generated sub σ-field of $\mathcal{B}$ which seperates points in E. Then $\mathcal{B} = \mathcal{B}_o$.

Now, we can formulate our result.

Proposition. Let $(L_\Phi, \mathcal{B}_{L_\Phi}, \mu)$ be a separable Orlicz space with the Borel σ-field $\mathcal{B}_{L_\Phi}$ and a probability measure μ. Let $\{\xi(t);\ t \in T\}$ be a measurable stochastic process on $(\Omega,\Sigma,P) = (L_\Phi, \mathcal{B}_{L_\Phi}, \mu)$ such that $\tilde{\xi}(x) = x$ a.e.. Then, there exists $N \in \mathcal{B}_{L_\Phi}$ $\mu(N) = 0$ such that the family $\{\xi(t);\ t \in T\}$ (restricted to $L_\Phi \setminus N$) generates $N' \cap \mathcal{B}_{L_\Phi}$ (in $N' \cap L_\Phi$).

Proof. Observe that $L_\Phi \subseteq L_0$, for every Φ. Moreover, L_Φ is a Borel set in L_0 (endowed with the Borel σ-field induced by the topology of convergence in measure m) and

$$\mathcal{B}_{L_\Phi} = L_\Phi \cap \mathcal{B}_{L_0}$$

So, we can assume without loss of generality, that $L_\Phi = L_0$.

Let $M = M(L_0,\mu)$ be the space of all (real) random variables on $(\Omega,\Sigma,P) = (L_0, \mathcal{B}_{L_0}, \mu)$ with the convergence in measure μ. ξ, in virtue of measurability induces the mapping $\bar{\xi}: T \longrightarrow M$

$$\bar{\xi}(t) = \xi(\cdot\ ,t) \in M.$$

$\bar{\xi}$ is a r.e. (with values in M) and has a separable range (Theorem 3), so it can be approximated (in the norm of M) by simple functions $\bar{\xi}^{(n)}$ of the form

$$\bar{\xi}^{(n)}(t) = \sum_{j=1}^{k_n} \xi(\cdot\ ,t_j^{(n)})\ \chi_{B_j^{(n)}}(t)$$

where $B_j^{(n)} \in \mathcal{T}$ and $t_j^{(n)} \in T$, $j = 1,\dots,k_n$; $n = 1,2,\dots$

By the measurability of $\xi^{(n)}$ it follows that we can assume that

$$\tilde{\xi}^{(n)}(x) \longrightarrow \tilde{\xi}(x) \quad \mu\text{-a.e.}$$

(we can choose a suitable subsequence n_k such that $\tilde{\xi}^{(n_k)}(x) \longrightarrow \tilde{\xi}(x)$ μ-a.e.). Since $\tilde{\xi}(x) = x$ μ-a.e., there exists $N \in \mathcal{B}_{L_0}$, $\mu(N) = 0$ such that if $x \notin N$ then $\tilde{\xi}(x) = x$ and $\tilde{\xi}^{(n)}(x) \longrightarrow x$

Let $\{\xi_j\}_{j=1}^{\infty}$ be a sequence containing all $\xi(\cdot\ ,t_j^{(n)})$, $j = 1,\dots,k_n$ $n = 1,2,\dots$. If $x,y \notin N$ and $x \neq y$ then there exists j such that $\xi_j(x) \neq \xi_j(y)$; for, if not, then

$$\tilde{\xi}^{(n)}(x) = \tilde{\xi}^{(n)}(y)$$

for every n, thus $x = y$, in virtue of the definition of N.

Thus, the countable family $\{\xi_j < r\}$, where r runs over all rationals and $j = 1,2,\ldots$, seperates the points of $N' \frown L_0$. By virtue of Blackwell's result, $\{\xi_j\}$ generates $N' \frown \mathcal{B}_{L_0}$.

Remark. If E is a separable Banach space then it is well known that the Borel σ-field in E is generated by all continuous linear functionals. Now, if ξ is a measurable stochastic process as described in Proposition, having the additional properties

$$\tilde{\xi}(x \pm y) = \tilde{\xi}(x) \pm \tilde{\xi}(y) \quad \mu \times \mu \text{ -a.e.}$$

(see Theorem 3) then $\xi(\cdot\ ,t)$ is an additive measurable functional on L_Φ, for almost all $t \in T$. Hence, we have the following : all additive measurable functionals obtained from a measurable process, with the properties as in Proposition, generate $\mathcal{B}_{L_\Phi}$ (mod μ).

Now, using Proposition we can easily prove the following result.

Theorem 5. Let ξ be a measurable Gaussian stochastic process with sample paths in a separable Orlicz space L_Φ. Then the measure μ_ξ induced by ξ on L_Φ is Gaussian.

Proof. We can assume that $L_\Phi = L_0$. Let us denote $\mu = \mu_\xi$, $\vartheta = 1/3\ \mu + 1/3\ \mu * \mu + 1/3\ \mu * \bar{\mu}$. Let η be a measurable process (see Theorem 3 and Proposition) on $L_0 \times T$ such that

$$\tilde{\eta}(x) = x \quad \vartheta\text{-a.e.}$$

and let $N \in \mathcal{B}_{L_0}$, $\vartheta(N) = 0$, be such that the family $\{\eta_t\}$ generates $N' \frown \mathcal{B}_{L_0}$. Since $\tilde{\eta}(x \pm y) = \tilde{\eta}(x) \pm \tilde{\eta}(y)$ $\mu \times \mu$ -a.e., $\eta(\cdot\ ,t)$ is a measurable additive functional, for $t \in T \setminus T_0$, $T_0 \in \mathcal{F}$, $m(T_0)=0$ We can assume that $T_0 = \emptyset$, without loss of generality.
Let

$$\zeta(\omega,t) = \eta(\xi(\omega),t).$$

Then ζ is a measurable process and $\tilde{\zeta} = \tilde{\xi}$ a.e.. By Fubini's Theorem, the random vectors

$$< \xi_{t_1},\ldots,\xi_{t_n} > = \ < \zeta_{t_1},\ldots,\zeta_{t_n} > \quad \text{a.e.},$$

for all $t_1,\ldots,t_n$ such that $t_i \notin T_1$, $i = 1,\ldots,n$, where $T_1 \in \mathcal{F}$, $m(T_1) = 0$. Putting $\eta_t' = 0$, if $t \in T_1$, and $\eta_t' = \eta_t$ if $t \notin T_1$, we obtain a measurable Gaussian process η', having all properties as η, so we can assume that $T_1 = \emptyset$.

Now, let $\{t_i\} = S$ be such that $\{\eta_{t_i}\}$ generate $N' \cap \mathcal{B}_{L_0}$ and let A_i, B_i be a finite collection of Borel subsets of the real line, $i = 1,\dots,n$. Denote

$$C = \bigcap_{i=1}^{n} \eta_{t_i}^{-1}(A_i), \quad D = \bigcap_{i=1}^{n} \eta_{t_i}^{-1}(B_i), \; t_i \in S.$$

Then, by the properties of η and in virtue of Bernstein's Theorem it is easy to check that

$$\mu \times \mu(\{(x,y) : x+y \in C, \quad x-y \in D\}) = \mu \times \mu(\{(x,y); \ x+y \in C\})$$
$$\mu \times \mu(\{(x,y); \ x-y \in D\}) = \vartheta_1(C) \ \vartheta_2(D)$$

where $\vartheta_1 = \mu * \mu$, $\vartheta_2 = \mu * \bar{\mu}$.

Since $\vartheta_1(N) = \vartheta_2(N) = 0$ so using standard arguments and the fact that the sets of the form C,D generates $N' \cap \mathcal{B}_{L_0}$, we obtain that this equality holds for all $C,D \in \mathcal{B}_{L_0}$, which completes the proof.

<u>Remark.</u> This theorem has been proved by Rajput [17], for L_p spaces, $1 \le p < \infty$. Observe that this theorem is valid also for measurable stochastic processes with values in LCA groups satisfying the second countability axiom. Another proof of this theorem, without using Blackwell's result, has been given in [5].

Theorem 4 and 5 establish a correspondence between Gaussian measures on L_Φ spaces and measurable Gaussian processes with sample paths in L_Φ. This correspondence allow us to solve many problems in these spaces. For example, given a Gaussian measure μ on L_0 and the corresponding measurable process ξ we can express the condition of the concentration of μ on L_Φ in terms of the covariance and the expectation of ξ (for some Φ's). We can also introduce in L_Φ spaces a notion of integral and of conditional expectation. Observe that, in general, these spaces need not be locally convex, so the classical notion of Bochner integral cannot be used. It is also possible to consider L_Φ-valued martingales and prove a mean convergence martingale theorem analogous to the theorem due to Chaterji [7], for Banach-valued martingales. As an application, we can obtain an orthogonal expansion of Gaussian r.e s with values in L_Φ spaces [6].

References

[1] C.R. Baker, Zero-one laws for Gaussian measures on Banach spaces, Trans. Amer. Math. Soc., 186 (1973), pp.291-308.

[2] P. Billingsley, Convergence of probability measures, New York 1968.

[3] D. Blackwell, On a class of probability spaces, Proceedings of the Third Berkeley Symp. on Math. Stat. and Prob., 1954-55, University of California Press, 1956, II, pp. 1-6.

[4] T. Byczkowski, The invariance principle for group-valued random variables, Studia Math., 56 (1976), pp. 187-198.

[5] -----, Gaussian measures on L_p spaces, $0 \leq p < \infty$, Studia Math., 59 (1977), pp. 249-261.

[6] -----, Norm convergent expansion for L_Φ-valued Gaussian random elements, to appear in Studia Math.

[7] S.D. Chaterji, A note on the convergence of Banach space-valued martingale, Math. Ann., 153 (1964), pp. 142-149.

[8] -----, Sur l'integrabilité de Pettis, Math. Zeitschrift, 136 (1974), pp. 53-58.

[9] K.L. Chung, J.L. Doob, Fields,optionality and measurability, Amer. J. Math., 87 (1965), pp. 397-424.

[10] L. Corvin, Generalized Gaussian measures and a functional equation I,J. Funct. Anal. 5 (1970), pp. 481-505.

[11] R.M. Dudley, M. Kanter, Zero-one laws for stable measures, Proc. Amer. Math. Soc., 45 (1974), pp. 245-252.

[12] X. Fernique, Integrabilité des vecteurs gaussiens, C.R., 270 (1970), pp. 1698-1699.

[13] M. Fréchet, Les éléments aléatoires de nature quelconque dans un espace distancie, Ann. Inst. H. Poincare, 10 fasc. 4.

[14] N.C. Jain, A zero-one laws for Gaussian processes, Proc.Amer. Math.Soc., 29 (1971), pp. 585-587.

[15] G. Kallianpur, Zero-one laws for Gaussian processes, Trans. Amer. math. Soc., 149 (1970), pp. 199-211.

[16] K.R. Parthasarathy, Probability measures on metric spaces, New York 1967.

[17] B.S. Rajput, Gaussian measures on L_p spaces, $1 \leq p \leq \infty$. J. Mult. Anal., 2 (1972), pp. 382-403.

[18] S. Rolewicz, Metric linear spaces, Warszawa 1972.

[19] D.S. Cohn, Measurable choice of limit points and the existence of separable and measurable processes, Z. Wahrscheinlichkeitstheorie verw. Geb., 22 (1972), pp. 161-165.

Institute of Mathematics
Technical University
50 - 370 Wrocław
Poland

SINGULARITY AND ABSOLUTE CONTINUITY OF MEASURES IN INFINITE DIMENSIONAL SPACES

S.D. Chatterji

§1. Introduction

In previous papers ([1(c)], [1(d)]), in collaboration with V.Mandrekar, we have developed a general technique for studying the singularity or absolute continuity of one probability measure with respect to another and have obtained as consequence the well-known results of Feldman, Hajek and others on Gaussian measures on the one hand and Kakutani's theorem concerning product measures on the other. We have used this latter theorem in [1(a)], [1(b)] to complete a result of Shepp. As pointed out in [1(a)], the problem could be studied for any measurable group G in place of R. The present paper takes a small step in this direction.

The general problem is as follows. Let G be a group endowed with a σ-algebra Σ such that multiplication ($(x,y) \rightarrow x\cdot y$) and inverse ($x \rightarrow x^{-1}$) are measurable operations. Such a pair (G,Σ) will be called a measurable group. Let μ be a probability measure on (G,Σ) and let P be the product probability measure $\mu \otimes \mu \otimes \ldots$ on the infinite product $G^\infty = G \times G \times \ldots$ endowed with the product σ-algebra $\Sigma^\infty = \Sigma \otimes \Sigma \otimes \ldots$. It is easily verified that $(G^\infty, \Sigma^\infty)$ is a measurable group. If $a = (a_n) \in G^\infty$, define P_a by $P_a(A) = P(A\cdot a)$, $A \in \Sigma^\infty$. Clearly $P_a = \otimes \mu_{a_n}$. The problem is to characterize the (P-quasi-invariance) set E(P) of all elements a in G^∞ such that P is equivalent to P_a (in symbols: $P \sim P_a$). The case $G = \mathbb{R}$ was treated in [1(b)]. In this paper, we shall settle the case of a general finite group G and point out some simple things in the case of a locally compact abelian group G. In particular, the results of [1(b)] go over verbatim to the groups $G = \mathbb{T}^k$ or $\mathbb{R}^k$. We hope to return to a more detailed study of other groups G in the near future.

It should be remarked that the following more general problem seems not to have been discussed systematically in the literature. Let (G,Σ) be a measurable group, μ a probability measure on G and E(μ) the μ-quasi-invariance subset in G. What can be said about

$E(\mu)$ as a subset of G? It is clear that $E(\mu)$ is always a subgroup of G. If G is a separable metric group then $E(\mu)$ can be proved to be a measurable subgroup of G. By varying μ, which measurable subgroups would one get? If G is a discrete (i.e. Σ = all subsets of G), denumerable group, then it can be easily seen that any subgroup can be realised as $E(\mu)$ for some probability μ. If G is a discrete non-denumerable group then a subgroup H of G is of the type $E(\mu)$ for some probability measure μ iff H is denumerable.

This can be seen as follows. Let p be the density of μ i.e. $\mu(A) = \sum_{x\in A} p(x)$. If $N = \{x \mid p(x) = 0\}$, $S = G\setminus N$ then S is denumerable and $E(\mu) = \{a \mid N = N\cdot a\} = \{a \mid S = S\cdot a\}$. Now if H is a denumerable subgroup and if we take S = H for some p, then for the corresponding probability measure μ, $E(\mu) = \{a \mid H = H\cdot a\} = H$. On the other hand, if μ is an arbitrary probability measure on G, then

$$E(\mu) = \{a \mid S = S\cdot a\} \subset F$$

where F is the subgroup generated by S. Since S is denumerable so is F and hence $E(\mu)$ is denumerable.

Of course, the known theorems of Weil-Mackey stating that $E(\mu) = G$ iff G is close to being locally compact (cf [2] p.52-53) are also interesting in this context.

§2. Preliminaries

We denote by e the unit element in G (by o if G is abelian in which case + would be used for group operation). By G^∞_F (resp. G^∞_o) we shall denote the set of all elements $a = (a_n) \in G^\infty$ such that $a_n \in F$ (resp. $a_n = e$) for n sufficiently large. In the sequel, μ will represent a probability measure on a measurable group (G,Σ) and $P = \mu \otimes \mu \otimes \cdots$ will represent the corresponding product measure on (G^∞,Σ^∞). In case G is locally compact, Σ will be the Borel σ-algebra of G. If μ is absolutely continous with respect to a Haar measure on G then we shall write $\mu(dx) = p(x)\,dx$ where dx represents Haar measure and $p = d\mu / dx$ is the density of μ with respect to Haar measure. For any measure μ on G, μ_t will be defined as the measure satisfying $\mu_t(A) = \mu(A\cdot t)$, $A \in \Sigma$, $t \in G$.

We note first that $P = \otimes\mu \sim P_a = \otimes \mu_{a_n}$ i.e. $a = (a_n) \in E(P)$ implies that $\mu \sim \mu_{a_n}$. Hence if G is locally compact and $E(P) \supset G^\infty_o$ (actually much less is required) then $\mu \sim \mu_t$ for all $t \in G$ whence

μ is equivalent to Haar measure on G. Thus $\mu(dx) = p(x)\,dx$ with $p(x) > o$ a.e. on G. Also $\mu_t(dx) = p(x\cdot t)\,dx$. In this case, by the above-mentioned theorem of Kakutani, we shall then have that $a = (a_n) \in E(P)$ iff

$$\prod_{n=1}^{\infty} \int_G \{p(x)p(x\cdot a_n)\}^{\frac{1}{2}}\,dx > o \tag{1}$$

Since

$$\int_G \{p^{\frac{1}{2}}(x) - p^{\frac{1}{2}}(x\cdot a_n)\}^2\,dx = 2\int_G \{1 - \sqrt{p(x)p(x\cdot a_n)}\}\,dx$$

we can re-write (1) as (putting $h = \sqrt{p}$)

$$\sum_{n=1}^{\infty} \int_G \{h(x) - h(x\cdot a_n)\}^2\,dx$$

$$= \int_G \sum_{n=1}^{\infty} \{h(x) - h(x\cdot a_n)\}^2\,dx < \infty\,. \tag{2}$$

The notations introduced explicitly or implicitly in sections §1 and 2 will be used henceforth without further explanation.

§3. G Finite

Theorem 1.

Let G be a finite group and μ be a probability measure on G having a strictly positive density. If P, E(P) are as above then $E(P) = G_F^{\infty}$ for some subgroup F of G. Conversely, if F is any subgroup of G then there is a probability measure μ with strictly positive density on G such that $E(P) = G_F^{\infty}$, $P = \mu \otimes \mu \otimes \cdots$

Proof: Let $p_1, p_2, \ldots$ be the distinct values of $p(x)$, $x \in G$ and let $S_j = \{x \in G \mid p(x) = p_j\}$. Then S_j's are disjoint sets and their reunion is G. From (2), we conclude that $a = (a_n) \in E(P)$ iff, for $n \geq N$, $a_n \in F$ where

$$F = \bigcap \{x^{-1} S_j \mid x \in S_j,\ j \geq 1\} \tag{3}$$

Clearly $e \in F$ and if α, β are in F then $\alpha\cdot\beta \in F$. Thus F is a finite semigroup in G whence we deduce that F is actually a subgroup. Thus $E(P) = G_F^{\infty}$. Conversely, given a subgroup F, we define a probability measure μ on G with strictly positive density p as follows: let $S_1 = F, S_2, S_3, \ldots$ be the distinct left cosets of F; define $p(x) = p_j$ for x in S_j in such a way that the p_j's are distinct, strictly positive and $\sum_x p(x) = 1$. From (3) it follows that for such a μ, $E(P) = G_F^\infty$.

Q.E.D.

For finite G, we can actually generalize the analysis above to cover the case of an arbitrary probability measure μ.

Theorem 2

Let μ be any probability measure on the finite group G. If P, E(P) are as above then $E(P) = H_F^\infty$ where $H = E(\mu) \subset G$, $F \subset H$ and H, F are subgroups of G. Conversely, given any two subgroups F and H with $F \subset H$, there exist probability measures μ on G such that $E(P) = H_F^\infty$ for the corresponding $P = \mu \otimes \mu \otimes \cdots$

Proof: Let p be the density of μ. If $a = (a_n) \in E(P)$ then, as pointed out before, $a_n \in E(\mu) = H$ for all n i.e. $a \in H^\infty$. As before, we see that $a = (a_n) \in E(P)$ iff $a_n \in H$ for all n and $a_n \in F_0$ for $n \geq N$ where F_0 is as in (3) but this time the S_j's correspond to the distinct non-zero values p_j of $p(x)$, $x \in G$. As before, F_0 is a subgroup of G. Thus $a_n \in H$ for all n and $a_n \in F = F_0 \cap H$ for $n \geq N$ iff $a = (a_n) \in E(P)$ i.e. $E(P)=H_F^\infty$. Conversely, given two subgroups $F \subset H$ we form a probability measure μ on G by defining its density p as follows: let $S_1 = F, S_2, \ldots$ be the distinct left cosets of F in H ; define $p(x)=p_j$ if $x \in S_j$ and $p(x) = o$ if $x \in G \setminus H$. Then $E(\mu) = H$ and by the reasoning above (cf (3)) we see that $E(P) = H_F^\infty$.

Q.E.D.

§4. G Discrete, Infinite

A complete analysis of this case can be carried out as in §3 but the results are more involved. For instance, if μ is an arbitrary probability measure on G then $a = (a_n) \in E(P)$, $P = \mu \otimes \mu \otimes \cdots$ implies that $a_n \in H = E(\mu)$ for all n and there exist a sequence of finite subgroups $F_j \subset H$ such that $a_n \in F_j$ for $n \geq N_j$. However, this latter more complicated condition does not suffice for $a = (a_n)$ to be in E(P). We postpone a general discussion to a later study, noting for the time being, the following easy generalisation of theorem 2.

Theorem 3

Let μ be an arbitrary probability measure on a discret group G. Suppose that G has no non-trivial elements of finite order. Then for $P = \mu \otimes \mu \otimes \cdots$, $E(P) = H_0^\infty$ where $H = E(\mu)$ is a countable subgroup of G. Conversely, for such a group G, any set H_0^∞ where H is a countable subgroup of G is E(P) for some $P = \mu \otimes \mu \otimes \cdots$

The proof of this is omitted since it follows easily from the previous discussions. We note in passing that the condition on G can be relaxed slightly by demanding that G have only a finite number of non-trivial finite subgroups.

§ 5. G Locally Compact Abelian

If μ is a Radon probability on G and $P = \mu \otimes \mu \otimes \cdots$ then $a = (a_n) \in E(P)$ only if $a_n \in E(\mu)$, $E(\mu) \subset G$, for all $n \geq 1$. If $\mu(dx) = p(x)\,dx$, $p(x) > 0$ a.e. on G then $E(P) \supset G_0^\infty$ (and conversely). We can reformulate (1) by passing to the Fourier transform $\hat{h}$ of $h = \sqrt{p}$. By Parseval's identity, we get from (1) that $a = (a_n) \in E(P)$ iff

$$\prod_{n=1}^{\infty} \int_{\hat{G}} |\,\hat{h}(\lambda)\,|^2\, \overline{\lambda(a_n)}\, d\lambda > 0$$

where $\hat{G}$ is the dual group of characters λ of G and $d\lambda$ is a Haar measure on $\hat{G}$ which matches dx on G in the Plancherel theory. Since

$$\int_{\hat{G}} |\,\hat{h}(\lambda)\,|^2\, d\lambda = \int_{G} |\,h(x)\,|^2\, dx = 1,$$

we see that $a = (a_n) \in E(P)$ iff

$$\sum_{n=1}^{\infty} \int_{\hat{G}} \{1 - \mathrm{Re}\,\lambda(a_n)\}\, |\,\hat{h}(\lambda)\,|^2\, d\lambda < \infty \tag{4}$$

A detailed knowledge of $\hat{G}$ permits us to study (4) in many significant cases. Thus, for $G = \mathbb{R}$, it is shown in [1(b)] that E(P) is a metrizable vector space of the type ℓ^ϕ (ϕ not necessarily convex) if E(P) is a vector space at all and that in general, E(P) is not a vector space (even if $G = \mathbb{R}$); a more explicit description of E(P) without any further hypotheses, even in this simple case, would seem to be unattainable. The analysis of [1(b)] repeated verbatim yields the following.

Theorem 4

Let G be T^k or $\mathbb{R}^k$ (where T = the circle i.e. the 1-dimensional torus group). Then if $\mu(dx) = p(x)\,dx$, $p(x) > 0$ a.e., $P = \mu \otimes \mu \otimes \ldots$ and if E(P) is such that $a \in E(P) \Rightarrow t\cdot a \in E(P)$ for any real t in some small interval around o then

$$E(P) = \{a = (a_n) \mid a_n \in \mathbb{R}^k, \sum_n \phi(a_n) < \infty\}$$

where

$$\phi(a) = \int_{\mathbb{R}^k} \min(|\lambda\cdot a|^2, 1)\ |\hat{h}(\lambda)|^2\, d\lambda\ , \quad a \in \mathbb{R}^k$$

or

$$\phi(a) = \sum_{\lambda\in Z^k} \min(|\lambda\cdot a|^2, 1)\ |\hat{h}(\lambda)|^2 \quad , \quad a \in T^k$$

$$h = \sqrt{p} \quad \text{and} \quad \lambda\cdot a = \lambda_1 a_1 + \ldots + \lambda_k a_k .$$

We hope to obtain further results of this type in the near future.

References

[1] Chatterji, S.D. and Mandrekar, V.

(a) Sur la quasi-invariance des mesures sous les translations. C.R.Acad.Sc. Paris 281, 581-583 (1975).

(b) Quasi-invariance of measures under translation. Math.Zeit. 154, 19-29 (1977).

(c) Singularity and absolute continuity of measures. Proceedings of the Conference on Functional Analysis, Paderborn, Germany (1976) Ed. K.-D. Bierstedt and B. Fuchssteiner, Notas de Matematica (63), 247-257 (1977).

(d) Equivalence and singularity of Gaussian measures and applications. In Probabilistic analysis and related topics, Ed. Bharucha-Reid, A.T. Academic Press (forthcoming).

[2] Varadarajan, V.S., Geometry of quantum theory, Vol.II, Van Nostrand (1970).

Dépt. de mathématiques
Ecole Polytechnique Fédérale
de Lausanne
61, av. de Cour
CH-1007 Lausanne
Switzerland

A COUNTER-EXAMPLE CONCERNING CLT IN BANACH SPACES

S.A. Čobanjan and V.I. Tarieladze

We discuss here conditions under which a measure defined on the Borel σ-algebra of a Banach space X satisfies the central limit theorem (CLT). If X is of type 2, then every measure with zero mean having strong second order satisfies CLT [2]; if X is of cotype 2, then every pregaussian measure satisfies CLT [6],[1]. Comparison of these facts leads us to the following question. Let ϑ be a measure on a Banach space X satisfying two conditions:

(a) ϑ is pregaussian,

(b) ϑ has strong second order.

Does ϑ satisfy CLT? We note that in aforementioned examples both these conditions are fulfilled: in spaces of type 2 (b) implies (a) and in spaces of cotype 2 (b) follows from (a). Let us remark that condition (a) is a natural necessary condition, while condition (b) is close to being a necessary one. The negative answer to this question is contained in [8], where an example is constructed of a pregaussian symmetric measure ϑ on C[0,1] concetrated on a unit ball (so ϑ has all strong orders), but not satisfying CLT. Recently Professor R.M. Dudley kindly informed the authors about the same example, when $X = c_0$. By extending of Dudley's idea we show that examples of these kind can be constructed for every Banach space containing l_∞^n uniformly. In particular there exist reflexive Banach spaces for which examples of this kind can be constructed.

Let X be a separable Banach space, X^* be the space of all continuous linear functionals on X, ϑ a Borel measure on X. We say that ϑ has weak p-th order ($p > 0$), if

$\int_X |\langle x^*, x\rangle|^p \vartheta(dx) < \infty$ for all $x^* \in X^*$ and ϑ has strong p-th order if $\int \|x\|^p \vartheta(dx) < \infty$. A measure γ on X is called Gaussian if any element $x^* \in X^*$ considered as a random variable on (X,γ) represents a (possibly degenerate) Gaussian random variable. A measure ϑ is called pregaussian, if it has weak second order and if there exists a Gaussian measure γ such that for each $x^* \in X^*$

$$\int < x^*, x >^2 \vartheta(dx) = \int < x^*, x >^2 \gamma (dx).$$

Let E_1, E_2 be isomorphic Banach spaces and $d(E_1,E_2)$ be the infimum of the number set $\{\|T\| \, \|T^{-1}\| \mid T: E_1 \longrightarrow E_2$ is an isomorphism $\}$. Denote R^n with the maximum-norm by l_∞^n. We shall say that a Banach space X contains l_∞^n uniformly, if for each $\varepsilon > 0$ and any integer n there exists an n-dimensional subspace E of X such that $d(E, l_\infty^n) \leq 1 + \varepsilon$. We shall need the following property of spaces containing l_∞^n uniformly [4], [7]: a Banach space X contains l_∞^n uniformly iff for each sequence (α_k) of positive numbers tending to zero there exists an unconditionally convergent series Σx_k, $(x_k) \subset X$ such that $\| x_k \| = \alpha_k$.

Let (Y_k) be a sequence of independent random elements with values in a separable Banach space, having the same distribution ϑ. We say that ϑ satisfies CLT if the sequence of normalized sums $n^{-1/2}(Y_1+\dots+Y_n)$ converges in distribution to a Gaussian measure with zero mean. It follows from the ordinary CLT that if ϑ satisfies CLT, then it is pregaussian and has zero mean. It is shown in [3] that if ϑ satisfies CLT, then ϑ has any strong p-th order, $0 < p < 2$, though ϑ does not necessarilly have strong second order.

__Theorem.__ __Let__ X __be a separable Banach space containing__ l_∞^n __uniformly. Then there exists a symmetric measure__ ϑ __on__ X __such that__

(i) __it is pregaussian, and__

(ii) __it is concetrated on a bounded subset of__ X , __but__

(iii) __it does not satify CLT.__

__Proof.__ Let us choose an unconditionally convergent series Σx_k, $x_k \in X$, such that $\|x_k\| = 1/\ln \ln \ln (k + 15)$, $k = 1, \dots$. We consider further a sequence of independent random variables (ξ_k) such that $P\{\xi_k = 1\} = P\{\xi_k = -1\} = 1/\ln(k + 7)$ and $P\{\xi_k = 0\} = 1 - 2/\ln(k + 7)$. Finally let us consider the random series

$$Y = \Sigma\, x_k\, \xi_k\,. \qquad (1)$$

The fact that the series $\Sigma\, x_k$ is unconditionally convergent implies that the series (1) is convergent a.s. and that its sum does not exceed $\sup\limits_{(\varepsilon_k),\ \varepsilon_k=\pm 1} \| \Sigma\, x_k\, \varepsilon_k \|$ a.s. Let ϑ be the distribution of the random element Y. We will show that ϑ is the desired measure. It is evident that ϑ satisfies (ii). Let us show that ϑ is pregaussian. We have

$$E < x^*, Y >^2 = \Sigma < x^*, x_k >^2\ 2/\ln(k+7).$$

But this quadratic form coinsides with that of the sum of the following a.s. convergent Gaussian series

$$\Sigma x_k\, (2/\ln(k+7))^{1/2}\ \gamma_k \qquad (2)$$

where (γ_k) is a sequence of independent standard Gaussian random variables. The series (2) is convergent since the sequence $(2/\ln(k+7))^{1/2}\ \gamma_k$ is bounded a.s. (cf.[9], p.72), while the series without this factor is unconditionally convergent.

Now we have only to show that ϑ does not does not satisfy CLT. Let (Y_k) be a sequence of independent identically distributed random elements with values in X having the distribution:

$$Y_1 = x_1\ \xi_{11} + \dots + x_k\, \xi_{1k} + \dots$$

$$\dots\dots\dots\dots\dots\dots\dots\dots\dots\dots\dots$$

$$Y_n = x_1\ \xi_{n1} + \dots + x_k\, \xi_{nk} + \dots$$

$$\dots\dots\dots\dots\dots\dots\dots\dots\dots\dots\dots$$

Here the elements of the matrix (ξ_{ij}) represent the jointly idependent system of random variables each ξ_{ij} having the same distribution as above ξ_j, $i = 1, \dots;$ $j = 1, \dots$.

Consider the following sets

$$E_{nk} = \bigcap_{i=1}^{n} \{\xi_{ik} = 1\}$$

It is clear that $P\{E_{nk}\} = \ln^{-n}(k+7)$. Let us introduce also the sets

$$A_n = \bigcup_{k<N_n} E_{nk} \text{, where } N_n = \exp\{\exp(n)\} .$$

we have

$$P\{A_n\} = 1 - P\{\bigcap_{k<N_n} E^c_{nk}\} = 1 - \prod_{k<N_n} (1-\ln^{-n}(k+7)).$$

Since

$$\prod_{k<N_n} (1-\ln^{-n}(k+7)) \leqslant (1-\ln^{-n}(\exp\{\exp\{n\}\}+7))^{\exp\{\exp\{n\}\}-1}$$

$$\sim (1-\exp\{-n^2\})^{\exp\{\exp\{n\}\}} \longrightarrow 0 \quad \text{when} \quad n \longrightarrow \infty$$

then $P\{A_n\} \longrightarrow 1$.

Let us write down $n^{-1/2} S_n$ in the form

$$n^{-1/2} S_n = \sum_k n^{-1/2} x_k \sum_{i=1}^{n} \xi_{ik}. \tag{3}$$

Assume now that $\omega \in A_n$. Then

$$\max_{1\leqslant k\leqslant N_n} \| n^{-1/2} x_k \sum_{i=1}^{n} \xi_{ik}(\omega) \| \geqslant \frac{n}{n^{1/2} \ln \ln \ln(\exp\{\exp\{n\}\}+15)} \sim$$

$$n^{1/2} \ln^{-1} n \longrightarrow \infty .$$

Now it is easy to deduce from this that the family $n^{-1/2} S_n$ is not bounded in probability (here we use the Levy inequality for the series (3) :

$$\underline{\lim}\, P\{n^{-1/2} \| S_n \| > n^{1/2} (2 \ln n)^{-1} \} \geqslant$$

$$2^{-1} \underline{\lim}\, P\{\sup_k \| n^{-1/2} x_k \sum_{i=1}^{n} \xi_{ik} \| > n^{1/2} \ln^{-1} n\} \geqslant$$

$$2^{-1} \underline{\lim}\, P\{A_n\} = 2^{-1} .$$

Remark : In the proof of Theorem we used only the existence in X of an unconditionally convergent series Σx_k such that $\| x_k \| = 1/\ln \ln \ln(k+15)$. However one can show (cf.[5]) that the existence of such series implies the existence of an unconditionally convergent series Σy_k such that $\| y_k \| = \alpha_k$, where (α_k) is any sequence of positive numbers tending to zero.

Therefore the class of spaces for which our construction goes through coinsides with that of spaces containing l_∞^n uniformly.

Corollary. There exist a separable reflexive Banach space X and a symmetric measure ϑ on X such that

(i) ϑ is pregaussian, and

(ii) ϑ is concentrated on a bounded subset of X, but

(iii) ϑ does not satisfy CLT.

As an example of a reflexive separable Banach space containing l_∞^n uniformly we can take $Z = \bigoplus_{k} l_\infty^n{}_{l_2}$. Elements of this space are all sequences (z_k), $z_k \in l_\infty^k$ such that $\Sigma \|z_k\|^2_{l_\infty^k} \leq \infty$; the corresponding norm on Z is $\|(z_k)\| = (\Sigma \|z_k\|^2_{l_\infty^k})^{1/2}$. It is evident that Z is reflexive and one can verify immediately that for every sequence (α_k) of positive numbers tending to zero there exists an unconditionally convergent series Σx_k such that $\alpha_k = \|x_k\|$. Consequently according to our Theorem the desired measure exists.

References

[1] Čobanjan, S.A. and Tarieladze, V.I., Gaussian characterizations of certain Banach spaces, J.Multivariate Analysis, 7 (1977), 183-203.

[2] Hofmann-Jørgensen, J., The strong law of large numbers and the central limit theorem in Banach spaces, Math.Inst.Aarhus Univ. Preprint Ser., 3 (1974/75).

[3] Jain, N.C., Central limit theorem in a Banach space, Lecture Notes in Mathematics, 526 (1975), 113-130.

[4] Maurey, B., Théorèmes de factorisation pour les opérateurs linéaires à valeurs dans les espaces L^p, Asterisque, 11 (1974), 1-163.

[5] Maurey, B. and Pisier, G., Caractérisation d'une classe d'espaces de Banach par des séries aléatoires vectorielles, C.R. Acad.Sci.Paris, 277 (1973), 687-690.

[6] Pisier, G., Le theoreme de la limite centrale et la loi du logarithme itere dans les espaces de Banach, Seminaire Maurey-Schwartz, 1975-1976, Exp. 3.

[7] Rakov, S.A., On Banach spaces for which Orlicz's theorem is not true (Russian), Matemat. Zametki, 14 (1973), 101-106.

[8] Strassen, V., and Dudley, R.M., The central limit theorem and ε-entropy, Lecture Notes in Mathematics, 89 (1969), 224-231.

[9] Vakhania, N., Probability distributions in linear spaces (Russian), Metzniereba, Tbilisi, 1971.

Academy of Sciences of the Georgian SSR
Computing Center, Tbilisi, 380093 USSR

RANDOM FUNCTIONS AND ORLICZ'S METHOD

REGULARITY OF PATHS AND LIMIT PROPERTIES .

X. Fernique

0. Introduction

The progress accomplished since fifteen years in the analysis of paths of gaussian random functions allowed since 1970 the study of some non-gaussian random functions. The typical case is this of sub-gaussian random functions where the majoration methods used in the gaussian case can be applied without modification as was shown by N.C. Jain and M.B. Marcus using covering technics and by B. Heinkel using integral bounds. In this methods and results, the functions $\exp x^2$ and $(\log \frac{1}{x})^{1/2}$ play an important role connected with the existence of exponential moments of the normal distribution. Another case related with the functions x^p has also been studied by M.G. Hahn using covering technics.

Our purpose here is to show that the results in random functions, gaussian or not, can be carried out using a unique method related to the structures of Orlicz spaces. This synthesis has been accomplished essentially by the work done in Strasbourg by N. Heinkel, C. Nanopoulos, Ph. Nobelis and the author.

1. Method of analysis.

Let (T,δ) be a separable metric space (not necessarily a Hausdorff space) whose diameter D is finite; B is the borelian subsets ; μ a probability measure on (T,B). Let Φ be a continuous convex function, with domain $\mathbb{R}_+$ satisfying:

$$\lim_{x\to 0}\frac{\Phi(x)}{x} = 0, \quad \lim_{x\to\infty}\frac{\Phi(x)}{x} = +\infty \ ;$$

let Ψ the conjugate function of Φ defined by :

$$\Psi(y) = \sup_{x\geq 0}\,[xy-\Phi(x)],$$

wich has the same properties as Φ . For any measurable complex valued function f defined on T, we set :

$$N_{\Phi,\mu}(f) = \inf \{\alpha > 0 : \int \Phi(|\frac{f}{\alpha}|)\, d\mu \leq 1\} ,$$

$$N_{\Psi,\mu}(f) = \inf \{\alpha > 0 : \int \Psi(|\frac{f}{\alpha}|)\, d\mu \leq 1\} ;$$

then we have :

$$\int |fg|\, d\mu \leq 2N_{\Phi,\mu}(f)N_{\Psi,\mu}(g).$$

The set $\mathcal{L}(\Phi,\mu)$ (resp. $\mathcal{L}(\Psi,\mu)$) of functions f for wich $N_{\Phi,\mu}(f)$ (resp. $N_{\Psi,\mu}(f)$) is finite, is a vector space and $N_{\Phi,\mu}$ (resp. $N_{\Psi,\mu}$) is a norm on it.

The procedure analysis of random functions that we present here can be summarize as following : Consider a random function X on T, wich is δ-separable, δ-separated and measurable, and a real sequence $(\varepsilon_n, n \in N)$ decreasing to zero. For every integer $n \geq 0$ and every $t \in T$, we set :

$$\mu_{n,t} = \mu\{s : \delta(s,t) \leq \varepsilon_n\} ,$$

$$\rho_{n,t}(s) = 0 \text{ if } \delta(s,t) > \varepsilon_n, \rho_{n,t}(s) = \frac{1}{\mu_{n,t}} \text{ if } \delta(s,t) \leq \varepsilon_n$$

$$g_{n,t}(u,v) = \sum_{k=1}^{n} [\rho_{k,t}(u)\, \rho_{k-1,t}(v) - \rho_{k-1,t}(u)\rho_{k,t}(v)],$$

$$g_t(u,v) = \lim_{n\to\infty} g_{n,t}(u,v) .$$

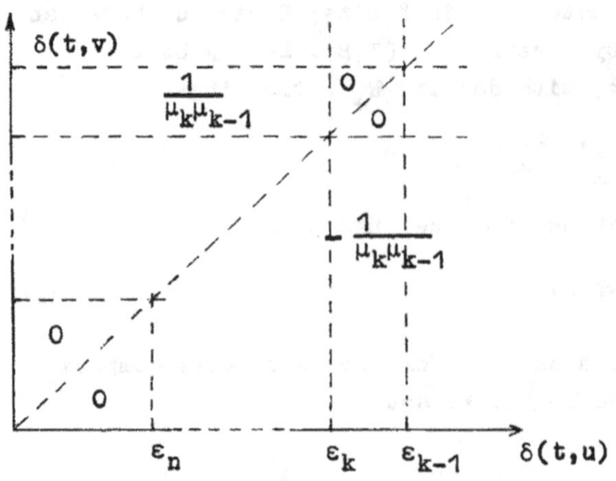

__Schema 1 : representation of variations of__ $g_{n,t}$ __when k vary from 1 to n.__

Suppose now that, for every $t \in T$, $\int_{\delta(s,t) \leq \varepsilon_n} X(\omega,s)\rho_{n,t}(s)d\mu(s)$ converge to $X(t)$ in probability; then we have $(\varepsilon_0 = D)$:

$$\forall \; t \in T, \; 2[X(t)- \int X(s)d\mu(s)] =$$

$$= \lim_{n\to\infty} \int\int_{\delta(u,v)\neq 0} [X(u)-X(v)]g_{n,t}(u,v)d\mu(u)d\mu(v) \;,$$

where limit is taken in probability. Let then ρ be a non negative increasing continuous function on $\mathbb{R}_+$; if $\frac{X(u) - X(v)}{\rho \circ \delta(u,v)} I_{\delta(u,v)\neq 0}$ belongs to $\mathcal{L}(\Phi,\mu \otimes \mu)$ a.s., and if $g_{n,t}(u,v)\rho \circ \delta(u,v)$, converges in $\mathcal{L}(\Psi,\mu\otimes\mu)$, we will obtain a.e. bounds of X and of his increments:

$$|X(t)- \int X(s)d\mu(s)| \leq N_{\Phi,\mu \otimes\mu}\left(\frac{X(u)-X(v)}{\rho \circ \delta(u,v)} I_{\delta(u,v)\neq 0}\right) \sup_{s\in T} N_{\Psi,\mu\otimes\mu}(g_s \, \rho \circ \delta)$$

and for every $\varepsilon > 0$ $(\varepsilon_0 = \varepsilon)$:

$$\sup_{\delta(t,t')\leq \varepsilon} |X(t) - X(t')| \leq$$

$$4N_{\Phi,\mu\otimes\mu}\left(\frac{X(u) - X(v)}{\rho \circ \delta(u,v)} I_{\delta(u,v)\neq 0}\right) \sup_{s\in T} N_{\Psi,\mu\otimes\mu}(g_s \, \rho \circ \delta) \, .$$

2. Examples of applications.

2.1. Sub-gaussian random functions: (N.C. Jain and M.B. Marcus, B. Heinkel).

We suppose that there exist two positive numbers a and b such that :

$$\forall \, (s,t) \;\; T \times T, \; E\{\exp a^2 \frac{|X(s) - X(t)|^2}{\delta^2(s,t)}\} \leq b \; ,$$

then we subgtitute to Φ the function $(e^{x^2} - 1)$; if $\rho(u) = u$ and $\varepsilon_k = \frac{\varepsilon_0}{2^k}$, we obtain :

$$|X(t) - \int X(s)d\mu(s)| \leq$$

$$KY(\omega) \sup_{s\in T} \int_0^D \sqrt{\log\left(1 + \frac{1}{\mu\{u : \delta(s,u)\leq x\}}\right)}dx$$

$$\sup_{\delta(t,t')<\varepsilon} |X(t) - X(t')| \leqslant$$

$$KY(\omega) \sup_{s\in T} \int_0^{\varepsilon} \sqrt{\log \left(1 + \frac{1}{\mu\{u: \delta(s,u) < x\}}\right)}\, dx$$

where Y is a random variable satisfying :

$$\forall x \geqslant \frac{1}{a}, \; P\{Y \geqslant x\} \leqslant b\, 2^{-a^2x^2} .$$

From the regularity of integral terms, we deduce the regularity of paths.

Studying the limit properties of X, we suppose that the following conditions is satisfied :

$$\forall \alpha \in \mathbb{R},\; \forall (s,t)\in T \times T,\; E\{\exp \alpha\, \mathcal{R}\left(\frac{X(s)-X(t)}{\delta(s,t)}\right)\} \leqslant e^{\frac{\alpha^2}{a^2}}$$

$$E\{\exp \alpha\, \mathcal{J}m \left(\frac{X(s)-X(t)}{\delta(s,t)}\right)\} \leqslant e^{\frac{\alpha^2}{a^2}} .$$

This condition implies that if S_n represents the sum of n independent random functions with the same distribution as X, we have :

$$\forall (s,t)\in T\times T,\; E\{\exp \frac{a^2}{16} \left|\frac{S_n(s)-S_n(t)}{\sqrt{n}\;\delta(s,t)}\right|^2 \} \leqslant 2.$$

The above computations imply that :

$$\forall n\in \mathbb{N},\; \forall \varepsilon > 0,\; E\{ \sup_{\delta(t,t')<\varepsilon} \left| \frac{S_n(t)-S_n(t')}{\sqrt{n}} \right| \} \leqslant$$

$$K' \sup_{s\in T} \int_0^{\varepsilon} \sqrt{\log\left(1 + \frac{1}{\mu\{\mu :\, \delta(s,u) \leqslant x\}}\right)}\, dx .$$

If the second member goes to zero with ε , then we deduce that X satisfies the central limit property.

2.2. Power type random functions (M.G. Hahn, C.Nanopoulos and Ph. Nobelis).

We suppose here the existence of a real $p > 1$ and a function f, increasing and continuous, such that :

$$\forall (s,t)\in[0,1]\times[0,1], E\{|X(s) - X(t)|^p\} \leq f^p(|s-t|),$$

by substituting to Φ the function x^p and setting, for technical reasons :

$$\rho(u) = u^{\frac{1+p}{p^2}} [f(u)]^{1-\frac{1}{p}}, \quad \varepsilon_k = \frac{\varepsilon_o}{2^k} ;$$

we obtain

$$E \iint_{[0,1]\times[0,1]} (|\frac{X(u)-X(v)}{\rho(|u-v|)}|^p) \, du \, dv \leq 2\int_0^1 \frac{f(u)}{u^{1+\frac{1}{p}}} du ,$$

from wich we deduce

$$\forall (s,t)\in[0,1]\times[0,1], |X(s)-X(t)| \leq KY(\omega)(\int_0^{|s-t|} \frac{f(u)}{u^{1+\frac{1}{p}}} du)^{\frac{p-1}{p}}$$

where Y is a random variable satisfying :

$$E(Y^p) \leq \int_0^1 \frac{f(u)}{u^{1+\frac{1}{p}}} du ,$$

and hence we obtain the regularity of paths if integral is finite.

In order to study the limit properties of X we will suppose $p \geq 2$ and $E[X^2(0)] < \infty$; this condition implies, for X centered (i.e. $\forall s \in [0,1]$, $E[X(s)] = 0$),

$$\forall n\in\mathbb{N}, \forall (s,t) \in [0,1]\times[0,1], E[| \frac{S_n(s)-S_n(t)}{\sqrt{n}} |^p] \leq K_p f^p(|s-t|)$$

where S_n is the sum of n independent random functions with same distribution as X. The previous computation implies that

$$\forall n \in \mathbb{N}, \varepsilon > 0, E\{ \sup_{|s-t|<\varepsilon} | \frac{S_n(s)-S_n(t)}{\sqrt{n}} |^p \} \leq$$

$$K_p \int_0^1 \frac{f(u)}{u^{1+\frac{1}{p}}} du \cdot (\int_0^{\varepsilon} \frac{f(u)}{u^{1+\frac{1}{p}}} du)^{p-1}$$

We deduce that X satisfies the central limit property if the integral converges in the neighbourhood of zero.

3. In conclusion, the present method of analysis has the advantage to reparate the variation of random functions in two parts, the random one and the temporal one ; the a.s. bounds given are easier to deal with than with the probability bounds.

References

[1] X. Fernique : Régularité des trajectoires des fonctions aléatoires gaussiennes. Lect. Notes Math., 480 (1975), pp. 1-96.

[2] M.G. Hahn : Conditions for sample continuity and central limit theorem. Ann. of probability, (1976), à paraître.

[3] B. Heinkel : Méthode des measures majorantes et théorême central limite dans C(S),Z. Wahrscheinlichkeitstheorie verw. Gebiete, (1977), à paraître.

[4] N.C. Jain et M.B. Marcus : Central limit theorem for C(S)-valued random variables. J. of Functional Anal., 19, (1975), pp. 216-231.

[5] C. Nanopoulos et Ph. Nobelis : Etude de la régularité des fonctions aléatoires et de leurs propriétés limites, Séminaire de probabilités de Strasbourg, 1977, à paraître.

Institut de Recherche Mathematique Avancee
Laboratoire Associé au C.N.R.S.
Université Louis Pasteur
7, rue René Descartes
67084 Strasbourg Cedex

F r a n c e

REMARKS ON POSITIVE DEFINITE OPERATOR VALUED FUNCTIONS IN LINEAR SPACES

Janusz Górniak

1. Introduction.

In this paper we present various applications of the dilations: (a) of positive definite functions with values in the space of all antilinear operators from a complex linear space to its algebraic dual ([6] Th.1) and (b) of positive definite functions with values in the space of all antilinear continuous operators from a complex locally convex space to its topological dual (see [5]) to analogues of Bochner's and Naimark's theorems.

Let X be a complex linear (locally convex) space with the algebraic (topological) dual $X'(X^*)$. By $\bar{L}(X,X')$ $(L(X,X^*))$ we denote the space of antilinear operators from X into $X'(X^*)$ and by $L(X,Y)$ - the space of all linear operators from X into a linear space Y.

For $A \in L(X,F)$, where F is a pre-Hilbert space with the inner product $[\cdot,\cdot]$ we define the algebraic adjoint operator $A' \in \bar{L}(F,X')$ by

$$(1.1) \qquad (A'f)(x) = [Ax,f], \qquad f \in F, \ x \in X.$$

Let Γ be a $*$-semigroup, i.e., Γ is a semigroup with the unit e and the involution "$*$" such that

$$\xi^{**} = \xi, \quad (\xi\eta)^* = \eta^*\xi^* \text{ and } e^* = e, \quad \xi,\eta \in \Gamma.$$

(1.2) Definition. Let $\mathfrak{X}$ denote a set of all functions $x = (x_\xi)$, $x: \Gamma \to X$ (having only a finite number of values different from zero). An $\bar{L}(X,X')$ - valued function $T = (T_\xi)$ on the $*$-semigroup Γ is called positive definite if

$$\sum_{\xi,\eta\in\Gamma} (T_{\xi^*\eta}\, x_\eta)(x_\xi) \geq 0$$

for each $(x_\xi) \in \mathfrak{X}$.

The function T is said to satisfy the <u>boundedness condition</u> if for each $\alpha \in \Gamma$ there exists $C_\alpha > 0$ such that

$$\sum_{\xi,\eta} (T_{\xi^*\alpha^*\alpha\eta}\, x_\eta)(x_\xi) \leq C_\alpha^2 \sum_{\xi,\eta} (T_{\xi^*\eta}\, x_\eta)(x_\xi)$$

for each $(x_\xi) \in \mathfrak{X}$.

We say that a family $D = (D_\xi)$, $D_\xi \in L(H,H)$ is a <u>representation of the *-semigroup</u> Γ in a Hilbert space H if D satisfies (1.4).

Now we give an analogue to Sz.-Nagy's dilation theorem ([7], p.458) in algebraic version.

(1.3) Theorem. ([6] Th.1). <u>Let</u> X <u>be a complex linear space with its algebraic dual</u> X'. <u>If an</u> $L(X,X')$ <u>- valued function</u> T <u>on</u> a *-semigroup Γ is positive definite on Γ, <u>then there exist</u> a <u>pre-Hilbert space</u> F <u>and a function</u> $D: \Gamma \to L(F,F)$ <u>with the properties:</u>

$$(1.4) \qquad D_e = I, \quad D_{\alpha\beta} = D_\alpha D_\beta \quad , \quad D_\alpha' = D_{\alpha^*} \, , \quad \alpha, \beta \in \Gamma$$

<u>and an operator</u> $A \in L(X,F)$ <u>such that</u>

$$T_\alpha = A' D_\alpha A \quad , \qquad \alpha \in \Gamma .$$

<u>Moreover,</u>

1° <u>the space</u> F <u>is minimal in the sense that it is generated by elements of the form</u> $D_\alpha Ax$, $x \in X$, $\alpha \in \Gamma$.

2° <u>the space</u> F <u>is unique up to unitary equivalence</u>, i.e., <u>if</u> $T_\alpha = A_1' D_\alpha^1 A_1$ <u>where</u> $D^1 : \Gamma \to L(F_1, F_1)$ <u>satisfies</u> (1.4), F_1 <u>is a minimal pre-Hilbert space (in the sense</u> 1° <u>) and</u> $A_1 \in L(X, F_1)$, <u>then there exists a unitary operator</u> $U: F_1 \to F$ <u>such that</u> $A = UA_1$ <u>and</u> $UD_\alpha^1 = D_\alpha U$.

3° <u>if</u> $T_{\xi\alpha\eta} = T_{\xi\beta\eta} + T_{\xi\gamma\eta}$ <u>for some fixed</u> $\alpha, \beta, \gamma \in \Gamma$ <u>and all</u> ξ, η <u>in</u> Γ <u>then</u> $D_\alpha = D_\beta + D_\gamma$.

In [6] the topological version of the theorem (1.3) in the case when X is a normed space is also provided. This theorem is true for a wider class of locally convex spaces. In [5] the large discussion is given about those locally convex spaces for which and only for which this theorem is true. Several new classes of locally convex spaces are presented and investigated (spaces with continuous Hilbert factorization property and s-factorization property, pseudo-barrelled spaces which contain plenty of well-known spaces: barrelled spaces, bornological spaces, quasi-barrelled spaces, DF-spaces, spaces with mixed topologies in Wiweger and Persson sense and some generalized inductive limits). For those kinds of spaces one provides analogues of Sz.-Nagy's dilation theorem ([7], p.458). We will present this theorem in the case when X is a barrelled space (barrelled spaces are included in all the above mentioned new classes of spaces). We will also give applications of this theorem for barrelled spaces.

(1.5) Theorem. Let X be a complex barrelled space with its topological dual X^*. If an $\bar{L}(X,X^*)$ - valued function T on a $*$-semigroup Γ is positive definite and satisfies the boundedness condition (see (1.2)), then there exist a Hilbert space H, a representation D of $*$-semigroup Γ in H and continuous linear operator $A : X \longrightarrow H$ such that

$$T_\alpha = A^* D_\alpha A \quad , \qquad \alpha \in \Gamma ,$$

where $A^* : H \longrightarrow X^*$ is the operator adjoint to A . Properties $1^0 - 3^0$ (from theorem (1.3)) hold and moreover,

4^0 $\|D_\alpha\| \leq C_\alpha$, $\alpha \in \Gamma$ (C_α as in the definition (1.2)),

5^0 if $T_{\xi\alpha_n\eta}$ (where n is from a directed set) converges to $T_{\xi\alpha\eta}$ in the weak operator topology of $\bar{L}(X,X^*)$ for all $\xi,\eta \in \Gamma$ and $\sup_n C_{\alpha_n} < \infty$, then $D_{\alpha_n} \longrightarrow D_\alpha$ in the weak operator topology of $\mathcal{L}(H)$.

Proof. The proof is analogue to the proof of the theorem 2 in [6]. We observe, that the operator $A: X \longrightarrow H$ is continuous. On the space X we define a family of seminorms $(p_y)_{y \in F}$, $\|y\| \leq 1$ (H is a completion of the pre-Hilbert space F) by

$$p_y(x) = |[y,Ax]| \ .$$

For each $y \in F$, $\|y\| \leq 1$, the seminorm p_y is continuous and the seminorm

$$p(x) = \|Ax\| = \sup_{\substack{y \in F \\ \|y\| \leq 1}} |[y,Ax]| = \sup_{\substack{y \in F \\ \|y\| \leq 1}} p_y(x)$$

is lower semicontinuous. Since X is a barrelled space, p(x) is continuous.

(1.6) Remark. The theorem (1.5) is true only in the case, when X is a locally convex space with continuous Hilbert factorization property (see [5]). Now we give an example of a locally convex space and a positive continuous operator $T \in \bar{L}(X,X^*)$ for which the theorem (1.5) is not true.

(1.7) Example. In the linear space s_0 we introduce a topology τ by the family of seminorms

$$p_{N_0,\{M_n\}}(x) = \sum_{n \in N_0} M_n |x_n| , \quad x = (x_1, x_2, \dots) ,$$

when N_0 is an arbitrary subset of natural numbers with the density equal to zero and $\{M_n\}$ - an arbitrary, non-negative sequençe of real numbers. The space (s_0,τ) doesn't have the factorization property (see [5]) and the operator $T: (s_0,\tau) \longrightarrow (s_0^*, \beta(s_0^*,s_0))$

$$(Tx)(y) = \sum_k \bar{x}_k y_k , \quad x,y \in s_0$$

is non-negative, continuous and doesn't have the factorization from the theorem (1.5) .

2. Bochner and Naimark type theorems.

Bochner's Theorem (see [4], § 8,Th.IV) has its analogue for positive definite $\mathcal{L}(H)$ - valued functions on a LCA group (see for example [4], §8,Th.VII). During the evolution of the theory of stochastic processes with values in Banach spaces ([8],[2],[3]) there was need to find an analogue to the Bochner Theorem for operator positive definite $\bar{L}(X,X^*)$ - valued functions, where X is Banach space - this theorem was given by Weron in [9] and [10].

(2.1) Definition. If X is a complex linear (locally convex) space, then the positive operator valued measure (P.O. measure) in X is a positive $\bar{L}(X,X')$ $(\bar{L}(X,X^*))$ - valued function K on the measurable space (S, Σ) such that $(K(.)x)(x)$ is a measure for all $x \in X$.

If X is a complex pre-Hilbert space and K_Δ $(\Delta \in \Sigma)$ is symmetric (i.e., $[K_\Delta x,y] = [x,K_\Delta y]$ for all $x,y \in X$), idempotent (i.e., $K_\Delta^2 = K_\Delta$) linear operator in X, and $K_S = I$, then we call such a P.O. measure a spectral measure in a pre-Hilbert space X.

A spectral measure K in a Hilbert space H is projection - valued P.O. measure.

Now we give an analogue to Bochner's Theorem for operator $\overline{L}(X,X')$ - valued (X is a linear space) functions on a LCA group. We start with the following remark :

(2.2) Remark. Stone's Theorem ([4],§8,Th.VI) is still valid if a Hilbert space H is replaced by a pre-Hilbert space F and a unitary representation U_g by a weakly continuous function $D_g (g \in G)$. The values of the function D_g are isometries onto the space F and they have property (1.4). Corresponding spectral measure usualy has its values in the completion $\widetilde{F}$ of F but not in F.

Proof. It's enough to use Stone's Theorem for the unitary representation $\{\widetilde{U}_g\}$ of the group G in a Hilbert space $\widetilde{F}$. The operators $\widetilde{U}_g$ are natural extension by the continuity of the operators D_g in F.

(2.3) Proposition. Let X be a complex linear space and $T = (T_g)$ be a weakly continuous $\overline{L}(X,X')$ - valued function on an LCA group G. If T is positive definite, then there exists a unique regular Borel P.O. measure (see def.(2.1)) K (in X) on the character group $\hat{G}$ such that for all $g \in G$

$$T_g = \int_{\hat{G}} < g,\gamma > \ K(d\gamma) \ , \qquad g \in G$$

(the integral being understood in the weak sense).

Proof. By theorem (1.3) there exist a pre-Hilbert space F, a function $D: G \longrightarrow L(F,F)$ with properties (1.4) and a linear operator $A: X \longrightarrow F$ such that

$$T_g = A' D_g A \ , \qquad g \in G \ .$$

For each $g \in G$ D_g is an isometry operator (see proof of theorem 2 in [6]) and it is a mapping F onto F, and function D_g is weakly continuous - the proof of that fact is the first part of the property 5° of theorem (1.5) - see [5]. By remark (2.2) there exists a unique regular spectral measure E with values in $\widetilde{F}$ on the group $\hat{G}$ such that

$$D_g = \int_{\hat{G}} < g,\gamma > E(d\gamma) \ , \qquad g \in G.$$

We extend the operator A' (described by (1.1)) to an operator A' (for simplicity we use the same letter) on $\widetilde{F}$ by

$$A'(\tilde{f})(x) = [Ax,\tilde{f}] , \quad x \in X, \quad \tilde{f} \in \tilde{F} .$$

It's obvious that A' is continuous when in X' we have the weak topology $\sigma(X', X)$.
Therefore

$$T_g = A'D_gA = \int_{\hat{G}} < g,\gamma > . A'E(d\gamma)A = \int_{\hat{G}} < g,\gamma > K(d\gamma),$$

where $K(.) = A'E(.)A$ is a P.O. measure.

One can formulate the next theorem (also (2.6),(2.7)) for locally convex spaces in the classes of spaces discussed in [5]. We'll give it for barrelled spaces.

(2.4) Proposition. Let X be a complex barrelled space and $T = (T_g)$ be a weakly continuous $\bar{L}(X,X^*)$ - valued function on an LCA group G. If T is positive definite then there exists a unique regular Borel P.O. measure K (in X) on the character group $\hat{G}$ such that

$$T_g = \int_{\hat{G}} < g,\gamma > K(d\gamma) \quad , \quad g \in G .$$

Proof. It's an analogue to the proof of Proposition (2.3) but we use Stone's Theorem and dilation theorem (1.5).

Naimark's Theorem about the factorization of $\bar{L}(X,X^*)$ - valued P.O. measures (where X is a Banach space) on $\mathbb{R}$ was given by Chobanyan in [1] (for the case of $\mathcal{L}(H)$ - valued measures see [4], 8, Th.VIII) and in the case of $\bar{L}(X,X^*)$ - valued P.O. measures on a measurable space (S, Σ) - by Weron (see [10], Proposition 3).

(2.5) Proposition. Let X be a complex linear space and $K(.)$ - $\bar{L}(X,X')$ - valued P.O. measure on the measurable space (S,Σ) . Then there exist a pre-Hilbert space F, a spectral measure E(.) in F on (S, Σ) and a linear operator $A: X \longrightarrow F$ such that for each $\Delta \in \Sigma$

$$K(\Delta) = A'E(\Delta)A .$$

Proof. The idea is like that in the proof of the Sz.-Nagy's dilation theorem ([7], p.463). Let us consider the σ - algebra Σ as a *-semigroup :

$$\Delta_1 \Delta_2 = \Delta_1 \cap \Delta_2 , \quad \Delta^* = \Delta , \quad e = S$$

and a P.O. measure is positive definite on this *-semigroup.
By theorem (1.3) there exists a pre-Hilbert space F, a function

$D: \Sigma \longrightarrow L(F,F)$ (with property (1.4)) and a linear operator $A : X \longrightarrow F$ such that

$$K(\Delta) = A'D(\Delta)A\ , \quad \Delta \in \Sigma\ .$$

Finally, we notice that $D(\Delta)$ ($\Delta \in \Sigma$) is a spectral measure in F. Therefore, for each $\Delta \in \Sigma$ operator $D(\Delta)$ is symmetric and idempotent and $D(e) = D(S) = I$ (it follows from properties $\Delta^* = \Delta$, $\Delta^2 = \Delta$, - see (1.4)) . σ-additivity of $D(\cdot)$ in the weak operator topology follows from property 5^0 of theorem (1.5) .

(2.6) Proposition. Let X be a complex barrelled space and $K(\cdot)$ - $\bar{L}(X,X^*)$ - valued P.O. measure on the measurable space (S,Σ). Then there exist a Hilbert space H, a spectral measure $E(\cdot)$ in H on (S, Σ) and continuous linear operator $A: X \longrightarrow H$ such that for each $\Delta \in \Sigma$

$$K(\Delta) = A^*E(\Delta)A\ .$$

Proof. The proof is similar to that of the Proposition (2.5) but we use theorem (1.5) (function $K(\Delta)$ on this same *-semigroup Σ is positive definite and satisfies the boundedness condition with constant $C_\Delta \equiv 1$).

Finally, one can obtain a factorization of P.O. measures in locally convex spaces by orthogonal measures. If X is a linear topological space and H is a Hilbert space then $L(X,H)$ - valued measure $M(\Delta)$ (i.e., $[M(\Delta)x,y]$ is a measure for all $x \in X$ and $y \in H$) on the measurable space (S, Σ) is orthogonal if $M^*(\Delta_1)M(\Delta_2) = 0$ for $\Delta_1,\Delta_2 \in \Sigma$, $\Delta_1 \cap \Delta_2 = \emptyset$.

(2.7) Proposition. Let X be a complex barrelled space and $K(\cdot)$ - $\bar{L}(X,X^*)$ - valued P.O. measure on the measurable space (S, Σ). Then there exist a unique minimal Hilbert space H and an orthogonal $L(X,H)$ - valued measure $M(\cdot)$ on (S, Σ) such that for each Δ Σ

$$K(\Delta) = M^*(\Delta)\, M(\Delta)\ .$$

Proof. Similar to that of Proposition 4 from [10] but we should use Proposition (2.6) .

(2.8) Remark. One can use the theorems given in this chapter to the characterization of a correlation function of X - valued processes where X is a locally convex space.

References

[1] Chobanyan, S.A., On some properties of positive operator valued measures in Banach spaces, (in Russian), Bull. Acad. Sci. Georg. SSR 57 (1970), 273-276.

[2] Chobanyan, S.A., Vakhania, N.N., Wide-sense valued stationary processes in Banach space, (in Russian), Bull. Acad.Sci.Georg. SSR 57 (1970), 545-548.

[3] Chobanyan, S.A., Weron, A., Banach space valued stationary processes and their linear prediction, Dissertationes Math. 125 (1975).

[4] Fillmore, P.A., Notes on operator theory, Van Nostrad Reinh. Math.Studies, vol. 30. 1970.

[5] Górniak, J., Dilations of locally convex spaces valued operator functions, (to be published).

[6] Górniak, J., Weron A., An analogue of Sz.-Nagy's dilation theorem, Bull.Acad. Polon.Sci., Ser. Sci. Math.Astronom. Phys. 24.10(1976), 867-872.

[7] Sz.-Nagy, B., Prolongement des transformations de l'espace de Hilbert qui sortent de ces espace, Appendix, F. Riesz et B. Sz.-Nagy, Leçons d'analyse fonctionelle, Paris-Budapest, 1965.

[8] Vakhania, N.N., Probability distributions on linear space, (in Russian), Mecnieraba, Tbilisi 1971.

[9] Weron, A., On positive definite operator valued functions in Banach spaces, (in Russian), Bull.Acad.Sci. Georg.SSR 71 (1973) 297-300.

[10] Weron A., Remarks on positive definite operator valued functions in Banach spaces, Bull. Acad. Polon. Sci., Ser. Sci. Math. Astronom. Phys. 24.10 (1976), 873-876.

Institute of Mathematics
Wrocław Technical University
50-370, Wrocław
P o l a n d

MIXTURES OF GAUSSIAN CYLINDER SET MEASURES AND ABSTRACT WIENER SPACES AS MODELS FOR DETECTION OF SIGNALS IMBEDDED IN NOISE

A.F. Gualtierotti

I. Introduction

We are going to consider some problems which are statistical in nature and arise when the variables considered take their values in spaces of functions. The modelling of complex systems, such as nonlinear filters, is often done as follows : first, some mathematical object is chosen as model not so much because it is believed to actually represent the system under investigation, but rather because it is the only one available. Typical is the modelling of noise with Gaussian statistics. Then, if numerical evaluations are needed, a reduction to a finite-dimensional model is conducted, and the ultimate decisions are based on the latter. Implicit in such a procedure are the following facts :

a) There is a connection between the finite-dimensional reduction and the original model.

b) The original model is "robust", for nobody expects it to be an exact representation of reality.

c) It is expected that known inadequacies of the model will not prevent one from learning something about the problem one is interested in.

Two related questions come to mind regarding the situation we have just sketched :

a) Is it possible to describe and understand mathematically the procedures mentioned above ?

b) Is it possible to incorporate intrinsically such procedures into a model ?

Indeed, there is a mathematical theory which in a sense deals with problems of this type, and that is the theory of abstract Wiener spaces; thus, we will try to use ideas from this theory to answer the questions we have asked above. But we shall only consider a particular problem, that of the detection of a signal at the end of a communication channel. The complete details for the results stated below will be given elsewhere.

II. The detection problem

A channel responds to a signal input by adding to it a noise. An observer at the end of the channel is given a sample function X and he has to decide whether he is given "noise only", or actually a distorted signal. Whatever the model for such a physical system, it has been agreed that if error-free decisions can be achieved through such a model, then the model is unworthy of further attention. The problem is then said to be SINGULAR. The models of interest are thus the NON-SINGULAR ones and these are mathematically the ones for which the following holds. Let (Ω, A, P) represent the system and X be the received waveform. For the "noise only" case (H_0) set $P_N = P \circ X^{-1}$, and for the "signal-plus-noise" case (H_1), set $P_{S+N} = P \circ X^{-1}$.
The problem is non-singular if $P_{S+N} \equiv P_N$. When it is established that the problem in non-singular, one can worry about statistical procedures such as likelihood inference. Typical problems are then the evaluation of Radon-Nykodim derivatives, and of the desired error probabilities.

III. Some models and some related problems

Because one deals with physical quantities, one takes the samples X to be in $L_2[0,T]$, $T < \infty$, hence the use of real and separable Hilbert spaces (denoted H). As to the probability laws involved, there is always a compromise between the necessity to compute and there necessity to model adequately. The models most dealt with are α) N and S+N Gaussian; β) N Wiener (or Poisson, or some "mixture"), S sufficiently smooth. We shall be concerned here with case α) only.

Let $B[H]$ denote the Borel sets of H, and P be a Gaussian measure on the Borel sets. P is characterized by its mean value m and its covariance operator R (trace-class, self-adjoint, positive). When m and R are known, there is a large amount of information available, and in the form one wants it, that is in terms of m and R.

When S is non-random a typical result is ; the detection is non-singular if and only if $S \in R^{1/2}H$. In practice, such results are inadequate for several reasons.

a) The condition $S \in R^{1/2}H$ is quite general, and thus difficult to check. Consequently, most of the known results are not readily applicable.

b) m and R are usually unknown and have to be estimated. Several difficulties immediately arise. For example :

  i) Do the estimates define Gaussian measures? In particular, if $\hat{R}$ is an estimate of R, is it trace-class ? In fact, estimates for covariance are few, their properties are not well known, and these estimates seem to perform poorly in detection problems.

  (ii) What relation does the detection problem based on the estimated parameters bear to the original problem? This question is all the more important that non-singularity is a fairly unstable "state" for such problems.

c) It has already been mentioned that a Gaussian model may not fit the physical systems studied adequately, and one needs to investigate how "robust" the involved decision procedures are, whether the parameters are estimated or not.

IV. The use of abstract Wiener spaces

One way to deal with the problems mentioned in III. is to accept whatever estimate one can put one's hand on and to search for a model which may accomodate this particular estimate. This is in sharp contrast with the usual procedures of statistics, where one wolud search, within the family of covariances, for the "best" estimator of, say, R. Such "ad hoc" methods may well be unavoidable Indeed, to check that an operators is trace-class, one needs an "infinite amount of data", for essentially one has to determine whether a series converges. So the critical element escapes observation, and no estimator that one can produce through observation can be included in the class one would, as a statistician, "naturally" search !

For the same reasons that one introduces the notion of non-singularity as a minimal requirement, we ask from any "ad hoc" model that it be consistent with the following "rules" :

a) What is "artificially" introduced into the problem should not be too different in nature from the ingredients of the problem at hand, and this for two reasons. The more "distance" there is between the original model and the "ad hoc" one, the more difficult it will be to justify the conclusions reached. For example, on a trivial sigma-algebra no problems is singular, but then a trivial sigma-algebra is not much of a model ! Furthermore, only on Hilbert spaces is the requested information available in terms of m and R (in the Gaussian case), and this we want to retain simply because it makes such measures so useful.

b) The decisions arrived at on the basis of "ad hoc" models should be less efficient, or of a narrower scope, that the decisions arrived at on the basis of "complete" information.

c) All the decisions should be arrived at on the basis of the original data, and not depend on what is introduced "a posteriori" into the problem.

The statistician has access to cylinder set measures, but needs to work with full measures, mostly because in infinite dimension, singularity problems are crucial. So he may "embed" his data into a larger space in such a way that these cylinder set measures he is working with become full measures. He can then take his decisions by inspecting the measures he has just constructed. This is exactly what is done in the theory of abstract Wiener spaces. But now we deal no longer with the standard Gaussian measure, but with a general Gaussian cyclinder set measure, obtained from a weak covariance operator (bounded, self-adjoint, positive), the estimate $\hat{R}$ of R.

Decisions thus arrived at will be "weak", in the same way that many estimates in statistics are "weakly" acceptable as soon as they converge weakly to the true value, for convergence almost surely cannot be achieved.

V. The model we consider

Let H be a real and separable Hilbert space, B[H] its Borel sets, and [H] be the family of bounded linear operators defined on H. We consider a map $f\colon \mathbb{R}_+ :=]0, \infty[\longrightarrow [H]$ such that $f(t) := R_t$ is a weak covariance operator, F-a.e. t, that is, R_t is a positive, self-adjoint and bounded injection of H, and F is a probability measure on the Borel sets of $\mathbb{R}_+$. We suppose further that $< R_t h,k >_H$ is measurable as a function of t, for every couple (h,k) in HxH, and also that $R_t h$, as a function of t, is Bochner integrable, for every h in H. One can then define a weak covariance operator R as follows :

$$Rh = \int_{\mathbb{R}_+} f(t)h \; dF(t).$$

Denote by μ^t the zero mean, Gaussian cylinder set measure determined by R_t.

The mixture μ is given by

$$\mu[C] = \int_{\mathbb{R}_+} \mu^t[C] \; dF(t),$$

for every cylinder set C, and its covariance structure is given by R. We shall suppose also that

A. $\int_{\mathbb{R}_+} ||R_t|| dF(t) < 0$, and

B. there exists a Borel set B of $\mathbb{R}_+$ with the following properties:

(i) $F(B) < 0$,

(ii) R_t has bounded inverse for t in B,

(iii) for some $c > 0$, $||R_t^{-1}|| \, ||R_t||^{1/2} < c^{-1}$, t in B.

Remarks:

1. On the use of mixtures

This is one way to deal with point c) of III, that is to get away from the Gaussian hypothesis. Mixtures can be used to model problems involving a randomness within the clock which monitors the system under observation. They are also natural models for multiple communication, where each element of the mixture describes the behaviour of one of the channels built in the system, and the mixing law says how access to the different channels is implemented. Finally, some simulated noise statistics could adequately by modelled with Gaussian mixtures.

2. On hypothesis A

This hypothesis in intuitively necessary to prevent μ from spreading out too much (that is the function of the trace-class property of covariance functions). Practically, hypothesis A is used to associate with the cylinder set measure μ a probability space, which plays a fundamental role in the development of the theory of abstract Wiener space.

3. On hypothesis B

This hypothesis says in particular that in some "neighborhood", μ^t behaves very much like the standard Gaussian distribution. As a consequence, the results presented have a shortcoming for essentially two reasons. Our implicit objective was to espace from constrains difficult to accomodate; thus hypothesis B goes againts such purposes. Still more unsatisfactory is the fact that the two most common procedures of estimation are excluded by the hypothesis we make ! Indeed, one method consists in estimating a "large" covariance matrix and treating the resulting finite dimensional problem. The second method, when X is stationary, consists in estimating the covariance of X and using this estimate as the kernel of the covariance operator $\hat{R}$. In neither case will bounded inverses of $\hat{R}$ exist. There are two admittedly not too consoling ways to live with hypothesis B: first, as already stated, the currently used estimation methods are not satisfactory either; and then one can view our case as a "limiting" case for the reduction method. Finally, it is hopefully worthwile to carry out the stated program, with a restrictive hypothesis, just to see what yields one gets !

In the development of the theory, hypothesis B is used to prove that measurable semi-norms are continous, and to construct a usefull family of finite dimensional projections and semi-norms.

4. A final point

The use of abstract Wiener spaces automatically takes care of reaquirement a) of section IV. We are thus left to check that requirements b) and c) can be accomodated as well, which then makes the method used worthy of further study, basically in order to free it from the unwieldy assumptions we have made.

VI. Some results

1. The basic tools

Definition [5; p.374]

A semi-norm q on H is μ-measurable if and only if, for every stricly positive ε, there is a finite dimensional projection Π_ε

such that, for any other finite dimensional projection Π orthogonal to Π_ε, one has

$$\mu[x \in H : q(\Pi x) > \varepsilon] < \varepsilon.$$

Such norms exist ($q(x) = ||Sx||$, S Hilbert-Schmidt) and are contінous, as a consequence of hypotheses A and B.

Proposition 1

Let q be a measurable semi-norm. Denote E_q the completion of H with respect to q and i_q the associated inclusion. Set $\mu_q = \mu \circ i_q^{-1}$: μ_q is a cylinder set measure on E_q.

Then μ_q extends to a probability measure μ_q on the Borel sets of E_q. Also $\mu_q^t = \mu^t \circ i_q^{-1}$ extends to a Gaussian probability measure $\tilde{\mu}_q^t$ on E_q, F-a.e. t, and $\tilde{\mu}_q$ is the mixture of the $\tilde{\mu}_q^t$'s for the mixing law F.

The proof is as in [7; p. 114, Thm. 1].

Proposition 2

H* is "the closure in $L_2[\tilde{\mu}_q]$ (resp. $L_2[\tilde{\mu}_q^t]$) of E_q^*.

Also $< \cdot, h >_H$ is a representative of the equivalence class of a random variable X_h with mean 0, variance $< Rh,h >_H$. (resp. $< R_t h,h >_H$) and the law of which is the mixture (normal N $(0, < R_t h,h >_H)$), of the normal laws $N(0, < R_t h,h >_H)$ with respect to F.

The proof is adapted from the standard Gaussian case [8; p. 78, Lemma 4.7].

Proposition 3

One can find a μ-measurable semi-norm q_0 and an increasing sequence Π_n of finite dimensional projections of H such that :

a) $id_H = \lim_n \Pi_n$ (strongly),

b) q_0 is stronger than q (hence $E_{q_0} \subseteq E_q$),

c) each Π_n extends by continuity to a projection $\Pi_n^{q_0}$ on E_{q_0},

d) $id_{E_{q_0}} = \lim_n \Pi_n^{q_0}$ (strongly).

The proof is as in [8; p. 66, Corollary 4.2], provided one uses the following lemmas when necessary :

Lemma 1

Let Φ_Σ be a Gaussian measure on $B[\mathbb{R}^n]$ with mean $\underline{0}$ and invertible covariance matrix Σ. Let T be a linear, symmetric and invertible operator on $\mathbb{R}^n$ such that $||T^{-1}||\ ||\Sigma^{-1/2}||\ ||\Sigma^{1/2}||\leq 1$. Then, if C is convex, centrally symmetric and closed,

$$\Phi_\Sigma[C] \leq \Phi_\Sigma[TC].$$

The proof uses [1 ; p. 184, Lemma (VI, 2;2)] after an appropriate change of variables is performed.

Lemma 2

Let H be a finite dimensional Hilbert space and μ_R be the Gaussian measure on H with mean 0 and covariance operator R. Let K be a vector subspace of H and C a convex, centrally symmetric and closed set.

Then, provided R is an injection,

$$\mu_R\,[C] \leq \mu_R\,[C \cap K+K^\perp].$$

The proof is as in [1 ; p. 186, Lemma (VI, 2;3)] provided one uses Lemma 1 when necessary.

2. Singular and non-singular detection for sure signals

Proposition 1

Let $\mathcal{R}_t$ denote the range of R_t and T_a the translation by a. Set $k = i_q(h)$ and suppose that the set of t's for which h is in $\mathcal{R}_t$ has F-measure one. Then

$$\tilde{\mu}_q \circ T_k^{-1} \equiv \tilde{\mu}_q \text{ (for some q).}$$

Also, if $\tilde{\mu}_q \circ T^{-1} \perp \tilde{\mu}_q$, the the set of t's for which h is in $\mathcal{R}_t$ has F-measure zero.

One can choose q as in Proposition 3 of 1. The translations of $\tilde{\mu}_q^t$ leading to equivalent measures are known to be given by the set

$$\left\{ \int_{E_q} f(x)\ x\ d\ \mu_q^t(x), \quad f\varepsilon\ H_t \right\} ,$$

where H_t is the closure of E_q^* in $L_2\ [\tilde{\mu}_q^t]$. One then shows, using Proposition 3 of 1, that

$$i_q \circ R_t(h) = \int_{E_q} f(x)x \ \ d\tilde{\mu}_q^t(x),$$

where $f(x) = \ < x,h >_H$, $\tilde{\mu}_q^t$ -a.s.

The result is established using then [9; p. 97, Corollary].

Remarks

Suppose $R_t = S \circ S_t$, where S_t has bounded inverse, for every t. Then $\mathcal{R}_t = \mathcal{R}(S)$, for every t. Consequently, the non-random signals leading to non-singular detection problems are in the range of S, rather than in the range of its square root. Since $\mathcal{R}(S) \subseteq \mathcal{R}(S^{1/2})$, Proposition 1 is thus in agreement with requirement b) of IV.

Proposition 2

Suppose that $||R_t^{1/2}|| \leqslant \beta < \infty$ and that $R_t^{1/2}$ has a bounded inverse F-a.e. t.

Then, for k not in $i_q(H)$, one has $\tilde{\mu}_q \circ T_k^{-1} \perp \tilde{\mu}_q$.

This result is a consequence of [4; p. 201, Thm. 1.2], when one knows that

$$\int_{E_q} q^2(x) \quad d\tilde{\mu}_q(x) < \infty .$$

The "square-integrability" of q follows from some inequalities to be found in [2; p. 903, Corollary 2.7; p. 909, iv; p. 913, Corollary 2.5] and Fernique's estimation [3]. Indeed these are used to check that

$$\sum_{n=1}^{\infty} \tilde{\mu}_q\ [q^2(x) > n] < \ \infty .$$

Remark

Proposition 2 is in agreement with requirement c) of IV. Elements outside of H to singular detection and may thus be ignored.

Proposition 3

Let R_0 be a weak covariance operator, with bounded inverse, determining a weak Gaussian distribution ν . For all $t > 0$, let S_t be a weak covariance operator, with bounded inverse, and such that $S_t - id_H$ is Hilbert-Schmidt. Suppose further that $R_t := R_0^{\frac{1}{2}} S_t R_0^{\frac{1}{2}}$, that $k = i_q(h)$ and that h is in the range of R_t, F-a.e. t.

Then

a) $\nu_q := \nu \circ i_q^{-1}$ extends to a probability measure $\tilde{\nu}_q$ on E_q ;

b) $\tilde{\mu}_q \circ T_k^{-1} = \tilde{\mu}_q$

c) $\frac{d}{d\tilde{\mu}_q} (\tilde{\mu}_q \circ T_h^{-1}) =$

$$\left\{\int_{R_+} \frac{d}{d\tilde{\nu}_q} (\tilde{\mu}_q^t) \, dF(t)\right\}^{-1} \left\{\int_{R_+} \frac{d}{d\tilde{\mu}_q^t} (\tilde{\mu}_q^t \circ T_k^{-1}) \frac{d}{d\tilde{\nu}_q} (\tilde{\mu}_q^t) \, dF(t)\right\} .$$

Fact a) is a consequence of [2; p.914, Corollary 2.7]. One then shows that $\tilde{\mu}_q^t = \tilde{\nu}_q$, by proving that, if $\tilde{R}_0$ and $\tilde{R}_t$ are the respective covariances of $\tilde{\nu}_q$ and $\tilde{\mu}_q^t$, one has

$$\eta(\tilde{R}_0\xi) - \eta(\tilde{R}_t\xi) =$$

$$\int_{E_q x E_q} q_t(w_1,w_2)\eta(w_1)\xi(w_2)d(\tilde{\nu}_q \otimes \tilde{\nu}_q)(w_1,w_2),$$

where $q_t \in L_2 \; [\tilde{\nu}_q \otimes \tilde{\nu}_q]$, η and ξ are in E_q^* . The latter formula is established using the properties of S_t. b) and c) then follow from Proposition 1 and [9; p.97, Lemma 1].

Remark

It may be of interest to notice that Proposition 3 is true for covariances having a certain form, here a product of operators. We may thus have a way to specify how the systems one works with should be built.

3. Non-singular detection for random signals

Proposition 1

Let R_t be invertible, F a.e. t, and T_t be a strongly measurable function of t such that, F-a.s.,

a) T_t is self-adjoint and Hilbert-Schmidt,

b) id_H+T_t is a weak covariance with bounded inverse.

Let ν^t be the weak Gaussian distribution associated with $S_t = R_t^{\frac{1}{2}}(id_H+T_t)R_t^{\frac{1}{2}}$ and set $\nu := \int_{\mathbb{R}_+} \nu^t \, dF(t)$.

Then ν_q extends to a probability measure $\tilde{\nu}_q$ on E_q which is equivalent to $\tilde{\mu}_q$.
The proof is similar to that of Proposition 3 of 2.

Proposition 2

Let $R := \sum_{i=1}^{\infty} \lambda_i e_i \circ e_i$, $0 < \alpha \leq \lambda_i \leq \beta < \infty$, and $S := R^{\frac{1}{2}}(id_H+T)R^{\frac{1}{2}}$, with id_H+T a weak covariance with bounded inverse and T trace-class.
μ denotes the weak Gaussian distribution associated with R.
Define A and B by $(id_H+T)^{\frac{1}{2}} = id_H+A$, $B = R^{\frac{1}{2}}(id_H+A)R^{-\frac{1}{2}}$.
Set $\nu = \mu \circ B^{-1}$. The ν is a weak Gaussian distribution with weak covariance S.
Define C, D and q as follows :

$$B^{-1} = id_H + R^{\frac{1}{2}}CR^{-\frac{1}{2}}, \quad R^{\frac{1}{2}}CR^{-\frac{1}{2}} = D, \quad q(x) = ||Dx||_H .$$

Then ν_q and μ_q extend to equivalent probability measures and

$$\frac{d}{d\mu_q}(\tilde{\nu}_q) = \det\{id_H+C\}\exp\{-\tfrac{1}{2}[2<CR^{-\frac{1}{2}}x,Rx^{-\frac{1}{2}}>_H + ||CR^{-\frac{1}{2}}x||_H^2]\},$$

where the right hand side is to be interpreted as a random variable on E_q.
The proof consists in a reduction of the problem to a finite dimensional subspace followed by a limiting procedure along the lines of [8; p.141, Thm.5.4].

Remark

This latter result is the prototype of what one would want in general. The final formula contains only ingredients derived from the original data and can be evaluated for any observation one actually comes up with [see IV.,c)].

It also confirms that what really matters is the relation of S to R [6].

Finally, Proposition 2 can be extended to cover mixtures also.

VII. Conclusion

If one is willing to forget under what restrictions the results of VI. are obtained, the program of III. and IV. has been carried out. To enable one to try to USE such results, the next step is to get rid of the restrictions made. But that has yet unfortunately, to be achieved!

References

[1] A.Badrikian and S.Chevet, Mesures cyclindriques, espaces de Wiener et fonctions aléatoires Gaussiennes, Lecture Notes in Math.379, Springer Verlag, Berlin/Heidelberg/New-York,1974.

[2] P.Baxendale, Gaussian measures on function spaces, Amer.J.Math., Vol.98(1976),pp.891-952.

[3] X.Fernique, Intégrabilité des vecteurs Gaussiens, C.R.Acad.Sci. Paris, Vol.270(1970), pp.A1698-9.

[4] A.Gleit and J.Zinn, Admissible and singular translates of measures on vector spaces, Trans.Amer.Math.Soc.,Vol.221(1976), pp.199-211.

[5] L.Gross, Measurable functions on Hilbert space, Trans.Amer. Math.Soc.,Vol.105(1962),pp.372-390.

[6] A.F.Gualtierotti, On the robustness of Gaussian detection, J.Math.Anal.Applic.,Vol.57(1977),pp.20-26.

[7] G.Kallianpur, Abstract Wiener processes and their reproducing kernel Hilbert spaces, Wahrscheinlichkeitstheorie verw.Geb. Vol.17(1971),pp.113-123.

[8] H.H.Kuo, Gaussian measures in Banach spaces, Lecture Notes in Math. 463, Springer Verlag, Berlin/Heidelberg/New-York,1975.

[9] A.V.Skorohod, Integration in Hilbert space, Springer Verlag, Berlin/Heidelberg/New-York,1974.

Dépt. de mathématiques
Ecole Polytechnique Fédérale
de Lausanne
61, av. de Cour
CH-1007 Lausanne
Switzerland

ON GAUSSIAN MEASURES AND THE CENTRAL LIMIT THEOREM IN CERTAIN F-SPACES

Werner E.Helm

Abstract. We present some new results concerning Gaussian measures on F-spaces with an absolute basis and apply these results to obtain a Central Limit Theorem for probability measures on arbitrary separable F-spaces.

1. Let E be a real linear space. A function $\|\cdot\| : E \longrightarrow [0,\infty]$ is called F-seminorm if it satisfies

   i) $\|x+y\| \leq \|x\| + \|y\|$, $\quad x,y \in E$

   ii) $\|\alpha x\| \leq \|x\|$, $\quad \alpha \in \mathbb{R}$, $|\alpha| \leq 1$, $x \in E$

   iii) $\|\alpha x\| \longrightarrow 0$, $\quad \alpha \in \mathbb{R}$, $x \in E$, $\alpha \longrightarrow 0$.

   $\|\cdot\|$ is an F-norm if $\|x\| = 0$ implies $x = 0$ and it is called p-homogeneous, $0 < p \leq 1$, if $\|\alpha x\| = |\alpha|^p \|x\|$ for $\alpha \in \mathbb{R}$, $x \in E$. Every F-norm defines a translation-invariant metric $d(x,y) = \|x-y\|$ which makes E into a linear metric space. If E is complete with respect to d(x,y) we call it F-space. We will consider separable F-spaces only. Provided with its Borel σ-algebra $\mathcal{L}$, $(E,\mathcal{L})$ then becomes a measurable linear space in the sense of Fernique [2], whose definition of a Gaussian measure we adopt. Let the mappings $T_{s,t} : E \times E \rightarrow E \times E$ be defined as

$$T_{s,t}(x,y) = (sx-ty,\ tx+sy), \qquad s,t \in \mathbb{R}.$$

We say that a probability measure γ on $(E,\mathcal{L})$ is mean zero Gaussian if the product measure $\gamma \otimes \gamma$ is invariant with respect to the family $\{T_{s,t} : s^2 + t^2 = 1\}$.
Concerning the integrability of F-norms with respect to Gaussian measures we have the following fundamental result.

<u>Theorem 1</u>. Let $(E,\|\cdot\|)$ be a separable F-space with p-homogeneous F-norm, $0 < p \leq 1$, let $r = 2/(2-p)$ and γ be a mean zero Gaussian measure on $(E,\mathcal{L})$. Then there exists an $\varepsilon > 0$ such that

$$\int \exp(\varepsilon \| x \|^r)\gamma(dx) < \infty .$$

Proof. The proof could be obtained through a very careful inspection and modification of Fernique's original proof [2]. However, using the previous result of Inglot/Weron [5], who have obtained Theorem 1 without the assumption of p-homogenity with $r = 1$, we may apply the following elementary idea (due to Hoffmann-Jørgensen).

Let $X^{(j)}$ be independent copies of X, $\mathcal{L}(X) = \gamma$, and set $S^{(n)} = X^{(1)} + \dots + X^{(n)}$. Then by definition, $\mathcal{L}(X) = \mathcal{L}(\frac{1}{\sqrt{n}} S^{(n)})$.

Let $a > 0$, $s = (2-p)/2$. An exponential Chebyshev estimate then yields

$$P(\| X \| > a\, n^s) \leq \mathbb{E} \exp\varepsilon \| S^{(n)} \| \exp(-\varepsilon a n^{s+ p/2})$$
$$\leq (\mathbb{E} \exp\varepsilon \| X \|)^n \exp(-\varepsilon a n).$$

Now we choose a so large to obtain, for some $\delta > 0$,

$$P(\| X \| > a\, n^s) \leq \exp(-\delta n) \qquad \text{for all} \quad n \in \mathbb{N}_0 .$$

Interpolating then yields the following tail probability estimation (C and β positive constants)

$$P(\| X \| > t) \leq C \exp(-\beta t^r) , \qquad t > 0 ,$$

and the proof follows simply by integration.

We remark that the above proof still works in the more general situation, when $(E,\mathcal{B})$ is just a measurable linear space and $\|\cdot\|$ a measurable pseudo-F-seminorm, which satisfies $\gamma(x: \|x\| < \infty) > 0$.

If E^*, the topological dual of E, separates points of E, our definition coincides with the usual one : γ is mean zero Gaussian if and only if every $f \in E^*$ is a (possibly degenerate) normally distributed real random variable with zero mean on $(E,\mathcal{B},\gamma)$. In this case we throughout assume all probability measures μ to be weakly centered, i.e.

$$\int f(x)\, \mu(dx) = 0, \qquad f \in E^* .$$

If μ has weak second moment, i.e. $E^* \subseteq L^2 \ (E,\mathcal{B},\mu)$, then

$$\Gamma_\mu(f,g) = \int f(x)g(x)\,\mu(dx), \qquad f,g \in E^*$$

will denote the covariance of μ , and we call μ pregaussian if it

has the same covariance as some Gaussian measure γ .
By $\mathcal{L}(X)$ we denote the distribution of a random element X defined on some probability space with values in E. Finally we say that μ satisfies the Central Limit Theorem (CLT) if the distributions of $\frac{1}{\sqrt{n}}(X^{(1)}+\ldots+X^{(n)})$ converge $\|\cdot\|$-weakly to a Gaussian measure γ on E, where $X^{(j)}$ are independent copies of X, $\mathcal{L}(X) = \mu$.

2. In [3] we had investigated Gaussian measures and the CLT in certain Orlicz sequence spaces determined by a sequence of subadditive Orlicz functions as follows

$$l_\phi = \{(y_j): y_j \in \mathbb{R}\ ,\ \sum_j \varphi_j(|y_j|) < \infty\}, \|y\|_\phi = \sum_j \varphi_j(|y_j|).$$

The interest in this class of sequence spaces containing all l^p-spaces, $0 \leq p \leq 1$ ($l^0 = \mathbb{R}^{\mathbb{N}}$) mainly stems from the following two propositions.

<u>Proposition 1</u>. Every F-space with an absolute basis is topologically isomorphic to a space l_ϕ .

<u>Proof</u>. First note that an F-space $(E,\|\cdot\|)$ has an absolute basis $\{e_j, e_j^*\}$ if $\Sigma_j \|e_j^*(x)\, e_j\| < \infty$ for every $x \in E$, $x = \Sigma_j e_j^*(x) e_j$.
Setting $\varphi_j(t) := \|t\, e_j\|$, $t > 0$, $j = 1,2,\ldots$ we obtain subadditive Orlicz functions in the above sense.
$u(x) = (e_j^*(x))_{j=1,2,\ldots}$ then defines a linear map $u : E \to l_\phi$ which is surjective because given $(\xi_j) \in l_\phi, u(z) = (\xi_j)$ if we put $z = \Sigma_j \xi_j e_j$ which is absolutely convergent in E.
Since

$$\|u(x)\|_\phi = \Sigma_j \varphi_j(|e_j^*(x)|) = \Sigma_j \|e_j^*(x)\, e_j\| \geq \|x\|$$

u has a continuous inverse, hence is a topological isomorphism by the open mapping theorem (applied to u^{-1}).

<u>Proposition 2</u>. [8] Every separable F-space is the quotient space of some l_ϕ-space.

<u>Proof</u>. We show that there exists a linear, continuous, open and surjective map $v: l_\phi \to E$. Let $\{a_j : j = 1,2,\ldots\}$, $a_j \neq 0$ be a countable dense subset of E, put $\varphi_j(t) = \|t\, a_j\|$, $t > 0$, $j = 1,2,\ldots$ and let l_ϕ be the

corresponding sequence space. If v is defined as $v((\xi_j)) = \Sigma_j \xi_j a_j$, then v is linear and continuous. Since, given $\varepsilon > 0$ we find $\delta > 0$ such that every $z \in E$, $\|z\| < \delta$ can be obtained as $\lim_n v(\xi^{(n)})$, $\xi^{(n)} \in l_\phi$, $\|\xi^{(n)}\|_\phi < \varepsilon$ v is nearly open, hence open and surjective by the generalized open mapping theorem [1,p.436] and the proposition is proved.

3. Now in the study of measures on sequence spaces a large role is played by the so called "standard deviation vector" $\sigma(\mu)$ to be defined coordinatewise as

$$(\sigma(\mu))_j = \Gamma_\mu(e_j^*, e_j^*)^{1/2} = \left(\int y_j^2 \mu(dy)\right)^{1/2}, \quad j=1,2,\ldots .$$

For our present purpose the results of [3] may be summarized as follows. Note that in view of Proposition 1 all subsequent results hold good in F-spaces with an absolute basis, and thus could have been formulated as well in this seemingly more general setting. However, the essentials become clearer by working in sequence spaces.

<u>Theorem 2</u>. i) A probability measure μ on l_ϕ is pregaussian if and only if $\sigma(\mu) \in l_\phi$.

ii) A probability measure μ on l_ϕ satisfies the CLT if and only if it is pregaussian.

The importance of the standard deviation vector is further stressed by the following results, the proofs of which we only outline.

<u>Theorem 3</u>. Let $\{\gamma_\alpha : \alpha \in A\}$ be a family of Gaussian measures on l_ϕ. Then $\{\gamma_\alpha : \alpha \in A\}$ is weakly relatively compact if and only if $\{\sigma(\gamma_\alpha) : \alpha \in A\}$ is relatively compact in l_ϕ.

<u>Proof</u>. The proof uses the relative compactness criterion for subsets of l_ϕ [3]. Sufficiency then follows through a slight strengthening of Theorem 4 of [3]. Since, clearly l_ϕ, and hence the set of all Borel probability measures on l_ϕ, too, are separable metric spaces, necessity follows by a standard subsequence-argument from the following lemma.

<u>Lemma 1</u>. [4]. Let $\gamma^{(n)}$ be a sequence of Gaussian measures on l_ϕ. If $\gamma^{(n)}$ converges weakly to a probability measure

μ on l_ϕ, then μ is Gaussian and $\sigma(\gamma^{(n)})$ converges to $\sigma(\mu)$ in l_ϕ.

Theorem 4. Let $X^{(i)}$ be a sequence of independent, weakly centered random elements with values in l_ϕ, $S^{(n)} = X^{(1)} + \dots + X^{(n)}$. If $\sigma(S^{(n)})$ converges in l_ϕ, then $S^{(n)}$ converges $\|\cdot\|_\phi$-almost sure.

Proof. Since for sums of independent random elements with values in separable F-spaces the three classic notions of convergence, almost sure convergence, convergence in probability and weak convergence coincide [7], it is sufficient to prove convergence in probability or weak convergence. Both can be done using methods developped already in [3] and [4]. For instance, using the inequality

$$\mathbb{E}\| S^{(m)} - S^{(n)}\|_\phi \leq K \cdot \|\sigma(S^{(m)} - S^{(n)})\|_\phi ,$$

it is not hard to show that $S^{(n)}$ is Cauchy, hence convergent in probability.

One may show by a counterexample that the above sufficient condition is strictly weaker than the one given by Woyczyński [9] for general F-spaces. In fact, if the random elements are Gaussian, it follows from Lemma 1, that the condition is necessary as well.

4. It is known that even in Banach spaces norm-conditions of the type $\int \|x\|^r \mu(dx) < \infty$ are not sufficient for μ to satisfy the CLT. Much less this can be expected in F-spaces, since there are F-spaces with bounded F-norm (e.g. L^0, the space of (equivalence classes of) all measurable functions on a finite measure space, provided with an F-norm inducing convergence in measure), where every probability measure has moments of all orders.
But the following general result holds. As before, $(E, \|\cdot\|)$ denotes a separable F-space.

Theorem 5. Let $X = \Sigma_i\, \eta_i(\omega) x_i$, $x_i \in E$, η_i real random variables, be a.s. absolutely convergent, $\mu = \mathcal{L}(X)$. If the variables η_i have zero expectations, finite second moment and

$$(*) \qquad \Sigma_i \|\sigma(\eta_i) x_i\| < \infty ,$$

then μ satisfies the Central Limit Theorem in E.

Proof. First observe that we do not assume the variables η_i to be independent and that, due to Proposition 2, every probability measure μ on E can be obtained as $\mu = \mathcal{L}(X)$ as indicated.

Setting as before $\varphi_j(t) = \|t\, x_j\|$, $t > 0$, $j=1,2,\dots$ and denoting by l_ϕ the corresponding Orlicz sequence space, $v : l_\phi \to E$, $v((\xi_j)) = \Sigma_j \xi_j x_j$ becomes a linear continuous map. Now $\eta(\omega) = (\eta_1(\omega), \eta_2(\omega), \dots)$ defines a random element with values in l_ϕ which is pregaussian and satisfies the CLT in l_ϕ because of Theorem 2 and condition (*). Hence there exists a Gaussian measure γ on l_ϕ such that $\theta_n(\vartheta^{n*})$ converges $\|\cdot\|_\phi$-weakly to γ, where $\theta_n x = (1/\sqrt{n})x$, ϑ is the distribution of η, ϑ^{n*} its n-fold convolution. Using the definition of a Gaussian measure it is easy to verify that $v(\gamma)$ is Gaussian on E and that $\theta_n(\mu^{n*})$ converges $\|\cdot\|$-weakly to $v(\gamma)$.

We remark that, due to its general nature, condition (*) is quite strong in many spaces. But according to Theorem 2, if $x_j = e_j$, it is necessary in all F-spaces with an absolute basis. Hence condition (*) cannot be weakened without restricting the class of spaces Theorem 5 is valid for.
In [5] Jain and Marcus prove a very neat CLT for random elements of a similar type with values in the Banach space C [0,1]. Theorem 5 clearly can be applied to obtain a CLT for such elements in the generally non locally convex Orlicz spaces

$$L_\varphi = \{f : [0,1] \to \mathbb{R} \;:\; \|f\| = \int \varphi(|f(t)|)dt < \infty\} ,$$

φ denoting a subadditive Orlicz function.

References:

[1] R.E.Edwards, Functional Analysis, Holt, Rinehart, Winston, 1965

[2] X.Fernique, Intégrabilité des vecteurs gaussiens, C.R.Acad. Sci. Paris 270 (1970), 1698-1699

[3] W.E.Helm, Gaussian Random Elements in Certain Orlicz Sequence Spaces, Bull.Acad.Polon.Sci. 25,5 (1977), 507-514

[4] W.E.Helm, Doctoral dissertation, TU Berlin

[5] T.Inglot,A.Weron, On Gaussian Random Elements in Some Non-Banach Spaces, Bull.Acad.Polon.Sci. 22 (1974), 1039-1043

[6] N.C.Jain,M.B.Marcus, Central Limit Theorems for C(S)-valued Random Variables, J.of Functional Anal. 19 (1975), 216-231

[7] A.Tortrat, Lois de probabilité sur un espace topologique complètement régulier et produits infinis à termes indépendants dans un groupe topologique, Ann.Inst.H.Poincaré 1,B (1965), 217-237

[8] Ph.Turpin, Convexités dans les espaces vectoriels topologiques généraux, Thèse, Orsay, 1974

[9] W.A.Woyczyński, Strong Laws of Large Numbers in Certain Linear Spaces, Ann.Inst.Fourier 24 (1974),205-223

FB Mathematik
TU Berlin
Strasse des 17, Juni 135
D-1000 B e r l i n 12 (WEST)

ON THE SPECTRAL MIXING THEOREM FOR SOME CLASSES OF BANACH SPACES AND FOR THE NUMERICAL CONTRACTIONS ON HILBERT SPACES

Vasile I.Istrătescu

0. Introduction. In [1] Blum and Hanson proved the following result for a mesure preserving transformation τ on a finite measure space $(\Omega, \mathcal{B}, m)$ and the induced unitary operator U on L^2 defined by Uf $(x) = f(\tau x)$: τ is strongly mixing if and only if

$$1/N \, \Sigma \, U^{k_i} f \rightarrow \int f(x)dm(x)$$

in the L^2 norm for all strictly increasing sequences $\{k_i\}$ and all $f \in L^2$. This result was extended in various directions and we mention here the following results obtained by Jones and Jones and Kuftinec [3],[4],[5]: the extension of the above result for a class of operators on a Banach space and to arbitrary contractions in Hilbert spaces.

In what follows we give a new proof of the spectral mixing theorem for contractions on Uniformly Convex Spaces using the theory of fixed points for nonexpansive mappings. This proof suggests extensions of the theorem to larger classes of Banach spaces for which we can apply some recent fixed point theorems.

Also using the notion of almost convergence introduced by G.G.Lorentz we consider some notions related to mixing and weak mixing as well as the analogue of the Mean Ergodic Theorem. In the last part we prove the Blum-Hanson theorem for a class of operators on Hilbert spaces larger than the class of contractions: the so called numerical contractions defined as all operators, such that

$$w(T) = \sup\{|\langle Tx,x \rangle| ; \|x\| = 1\} \leq 1.$$

1. The spectral mixing theorem for uniformly convex spaces.

Let X be a complex Banach space and T be a bounded linear operator on X. The operator T is called weak mixing to zero if

$$\lim 1/N \sum_{1}^{N} |x^*(T^n x)| = 0 \quad \text{for all } x \in X \text{ and } x^* \in X^*$$

and T is called strong mixing to zero if weak-lim $T^n x = 0$ for all $x \in X$. A famous result of von Neumann is the following : If U is a unitary operator on L^2 then T is weak mixing to zero iff U has no eigenvalues on the unit circle.

This result was extended to uniformly convex spaces by K. L.Jones [4] and the result is as follows : if T is a contraction on an uniformly convex space then T is weak mixing to zero if and only if T has ho eigenvalues of modulus one. The proof of this theorem uses some deep results of H.Weyl [11] on almost periodic vectors. We give now a proof using some fixed point theorems.

Definition 1.1. A mapping $f : C \to C$, where C is a closed convex set in a Banach space is called nonexpansive if for all $x,y \in C$, $\| f(x)-f(y)\| \leq \| x-y\|$.

The following theorem was proved by many authors, see [10] for details.

Theorem 1.2. If f is a nonexpansive mapping of a closed convex and bounded set in a uniformly convex space then f has a fixed point in that set.

Now we are ready to prove the spectral mixing theorem.

First we remark that if for some $x_0 \neq 0$, $Tx_0 = z_0 x_0$ and $|z_0| = 1$ then T is not weak mixing to zero.

Conversely, suppose now that T is not weak mixing to zero and we prove the existence of an eigenvalue of modulus one.

We find a nonzero recurrent vector, say x_0. From the definition of recurrent vectors it follows that there exists a sequence of integers, $\{n_i\}$ such that $T^{n_i} x_0 \to x_0$ and since T is a contraction, we have $\| Tx_0\| = \| x_0\| = \| T^i x_0\|$ for all $i \geq 1$.

Let

$$y = \sum_{1}^{N} \alpha_i T^i x_0 , \alpha_i \in C .$$

Since

$$\| T^k y-y\| = \| \sum_{1}^{N} \alpha_i T^i(T^k x_0 - x_0)\| \leq (\sum_{1}^{N} |\alpha_i|)\| T^k x_0 - x_0\|$$

we obtain that y is again a recurrent vector. As above we obtain that $\| T^i y\| = \|y\|$ for all $i \geq 1$. Consider now the following set

$$C = \{y : y = \sum_1^N \alpha_i T^i x_0, \alpha_i \geq 0 , \sum_1^N \alpha_i = 1\}$$

which is obviously a convex and bounded set in X. Also from the definition C is invariant for T. Another important property of T regarding the set C is that T is nonexpansive on C. It is clear that all the above properties extend to the closure of C, denoted by Cl C. From the above theorem about fixed points we obtain that there exists a point y^* in Cl C such that $Ty^* = y^*$. It is clear from the definition of C that $y^* \neq 0$.

This result proves the theorem.

The proof given above admits extension to the case of some Banach spaces for which the closed convex and bounded sets are with some special properties.

First we consider the case of Banach spaces with the so called normal structure. A subset M of a Banach space has normal structure [7] if for each bounded convex subset C of M which contains more than one point there exists a point $x \in C$ which is not a diametral point of C (that is, $\sup\{\| x-y\| , y \in C\} < \text{diam } C$).

We say that a Banach space has a normal structure iff any subset of it has a normal structure.

Suppose that X is a Banach space with the following properties:

1. X is reflexive,
2. X has a normal structure,
3. if T is a contraction which is not weak mixing to zero then T has nonzero recurrent vectors.

We have the following result:

<u>Proposition 1.3</u>. If X is a Banach space having the above three properties then any contraction T is weak mixing to zero iff T has no eigenvalue of modulus one.

Proof. If T has an eigenvalue of modulus one we obtain easy that T is not weak mixing to zero.

Conversely, if T is not weak mixing to zero we can construct the set C as above and for the set Cl C we can apply the fixed point of Kirk [6] and we obtain the existence of an eigenvalue of modulus one. This proves the Proposition.

For the next result we need the notion of essential spectral radius of an operator $T \in L(X)$. As is well known there are several definitions of the essential spectrum of an operator $T\varepsilon L(X)$.

The definitions which result in the smallest set is due to Kato [14] who defines $\sigma_e(T)$ to be the complement of the set $F_T = \{z \; ; \; T-z$ is a semi Fredholm operator$\}$ and the definition which results in the largest set is due to Browder [15] who defines $\sigma_e(T)$ as the complement of the union of all components containing points of the resolvent. All other definitions of the essential spectrum fall between those of Kato and Browder. The following formula for the calculus of the essential spectral radius (for any definition of the essential spectrum) is as follows:

$$r_{ess\ T} = \lim \alpha(T^n)^{1/n}$$

where for any operator $S\varepsilon L(X)$, $\alpha(S)$ is the inf $\{a; a>0, S$ is a a-set contraction$\}$.

Our result is the following:

<u>Proposition 1.4</u>. If T is in L(X), has uniformly bounded iterates and $r_{ess\ T}$ is less than 1 then if X satisfies the property 3 then T is weak mixing to zero iff T has no eigenvalues of modulus one.

Proof. As above we see that the condition is necessary.

Conversely, if T is not weak mixing to zero then we show that T has an eigenvalue of modulus one. Indeed, since T has a nonzero recurrent vector, say x_0, as in the case of uniformly convex spaces we can construct the set C which is invariant for T.

Also since T has the $r_{ess\ T}$ less than one, using an idea of Krasnoselskii [8] we can define a new norm on X equivalent with the original norm of X and such that T is an $r_{ess\ T} + \varepsilon$-set contraction.

In this case, by Darbo's generalization of the Schauder fixed point theorem, there exists a point y_0 in Cl C such that $Ty_0 = y_0$. This proves the assertion.

2. Almost mixing and almost ergodicity.

In what follows we use the idea of G.G.Lorentz about almost convergence to consider an extension of the notion of ergodicity, weak ergodicity etc.

First we give the corresponding definitions.

Let X be a Banach space and $T \in L(X)$ uniformly bounded iterates, that is there exists $M < \infty$ such that for all n, $\|T^n\| \leq M$.

Definition 2.1. The operator T is called almost ergodic at x if there exists $x_0 \in X$ such that

$$\lim_p (T^n x + T^{n+1} x + \ldots T^{n+p} x)/p = x_0$$

uniformly in n.

Definition 2.2. A sequence $\{x_n\}$ in X is called weak almost convergent to x_0 if for all $x^* \in X^*$, $\{x^*(x_n)\}$ is almost convergent to $x^*(x_0)$, i.e.,

$$\lim_p \{(x^*(x_n) + x^*(x_{n+1}) + \ldots + x^*(x_{n+p})\}/p = x^*(x_0)$$

uniformly in n for each $x^* \in X^*$.

Definition 2.3. The operator $T \in L(X)$ is called almost strong mixing at x if $\{T^n x\}$ is weak almost convergent to $x_0 \in X$.

As is well known the famous Mean Ergodic Theorem asserts that if for a given $x \in X$ there exists a subsequence of the sequence

$$\{1/n \sum_{0}^{n} T^i x\}$$

weakly convergent to x_0 then T is ergodic at x and

$$\lim 1/n \sum_{1}^{n} T^i x = x_0 .$$

First we give the extension of the Mean Ergodic Theorem for the case considered by us.

Theorem 2.4. Let $T \in L(X)$. If for some subsequence $\{n_k\}$,

$$\{1/n_k \sum_{1}^{n_k} T^i x\}$$

is weakly almost convergent then T is almost ergodic at x and

$$Tx_0 = x_0 .$$

Proof. The proof is a slight modification of the proof of The Mean Ergodic Theorem and thus we omit it.

Corollary 2.5. If T is almost strong mixing at x then T is amost ergodic at x.

The following result gives a connection between weak convergence and the almost convergence.

Theorem 2.6. Let T be a contraction on a Hilbert space. Then for any $x \in H$, $T^n x \rightharpoonup x_0$ iff $\{T^{k_i} x\}$ is almost strongly convergent for all strictly increasing sequences of positive integers.

Proof. The necessity is as in the Theorem [5] of Jones and Kuftinec. For the converse we use also the method of Jones and Kuftinec.

Suppose that $T^n x \rightharpoonup x_0$ and since x_0 is a fixed point of T we may suppose without loss of generality that $x_0 = 0$.

Also we may assume that $1 = \inf \|T^n x\|$.

Let $\varepsilon > 0$ and we find the integers M and K such that

$$1.\ \| T^M x\|^2 \leq 1+\varepsilon ,$$

$$2.\ |< T^M x, T^{M+k} x >| < \varepsilon \text{ for } k > K.$$

As in [5] we obtain that for $m > M$, $n > M$

$$\mathrm{Re} < T^m x, T^n x > \leq 4\varepsilon .$$

Suppose now that $\{k_i\}$ is a strictly increasing sequence of positive integers, and thus we have

$$\| 1/p \sum_{i=1}^{p} T^{k_i+n} x\|^2 = \mathrm{Re} < 1/p \sum_{i=1}^{p} T^{k_i+n} x,\ 1/p \sum_{i=1}^{p} T^{k_i+n} x > =$$

$$= 1/p^2 \sum_{i,j} \mathrm{Re} < T^{k_i+n} x,\ T^{k_j+n} x > =$$

$$= 1/p^2 \sum_{i=1}^{p} \| T^{k_i+n} x\|^2 + 2/p^2 \sum_{\substack{i<M \\ i<j}}^{p} \mathrm{Re} < T^{k_i+n} x,\ T^{k_j+n} x > +$$

$$+ 2/p^2 \sum_{\substack{i\geq M \\ i<j<K+i}} \mathrm{Re} < T^{k_{i+n}}x, T^{k_{j+n}}x > + 2/p^2 \sum_{\substack{i\geq M \\ j\geq K+i}} \mathrm{Re} < T^{k_{i+n}}x, T^{k_{j+n}}x >$$

$$\leq \|x\|^2/p^2 + 2(M-1)p\,\|x\|^2/p^2 + 2(K-1)p\|x\|^2/p^2 + 4\varepsilon$$

which for p large is small uniformly in n; hence the theorem holds.

3. Blum-Hanson theorem for numerical contractions.

We recall that an operator T on a Hilbert space is called a numerical contraction if

$$w(T) = \sup\{|< Tx,x >|, \|x\| = 1\} \leq 1.$$

Also, another useful notion is that of dilation of operators.

If T is an operator on a Hilbert space H, we say that $T^{\sim}$ is the dilation of T if:

1. $T^{\sim} \in L(K)$, K is a Hilbert space containing H as a closed subspace, and

2. for any n, $T^n x = PT^{\sim n}Px$.

As is well known any contraction has a unitary dilation, i.e. the operator $T^{\sim}$ can be chosen unitary.

An important and useful result of C.Berger is that any numerical contraction T satisfies

$$T^n x = 2PU^nPx \;, \qquad n \geq 1,$$

where U is a unitary operator defined on a Hilbert space K as above and P is the othogonal projection of K onto H.

From this structure theorem we show that we can obtain the Blum-Hanson theorem for contractions and for numerical contractions.

First we remark that the following lemma holds.

Lemma 3.1. Let H be a Hilbert space, P a Hermitian projection and U a unitary operator such that for $x \in H$, $PU^nPx \to 0$.
Then

$$1/n \,\Sigma\, PU^nPx \to 0.$$

Proof. The proof can be modelled on the proof of [5] and thus we omit it.

As a consequence we obtain,

Theorem 3.2. If T is a numerical contraction on a Hilbert space then for any $x \in H$, $T^n x \to x_0$ iff $1/N \sum_{i=1}^{N} T^{k_i} x \to x_0$ for all strictly increasing sequences $\{k_i\}$ of positive integers.

Proof. The necessity follows as in the case of contractions and the assertion in the oposite direction follows from the above Lemma.

Remark 3.3. The above proof suggests the possibility that a theorem of Blum-Hanson type can hold for other classes of operators on Hilbert spaces [9].

The result about numerical contractions was obtained with Ana I.Istrăţescu. The author thanks for discussions with A.Istrăţescu about these problems as well as for the permission to include the above result here.

References

[1] J.R.Blum and D.L.Hanson, On mean ergodic theorem for subsequences. Bull.A.M.S.66(1960) 308-311.

[2] G.Bennet and N.J.Kalton, Consistency theorems for almost convergence. Trans.A.M.S.198(1974) 23-43.

[3] L.K.Jones, A Mean Ergodic Theorem for Weakly Mixing Operators. Adv.in Mat. 7(1971) 211-216.

[4] L.K.Jones, An elementary lemma on sequences of integers and its applications to functional analysis. Math.Z.126(1972) 299-307.

[5] L.K.Jones and V.Kuftinec, A note on the Blum-Hanson theorem Proc.A.M.S.30(1971) 202-203.

[6] W.A.Kirk, A fixed point theorem for mappings which do not increase distances. Amer.Math.Month.72(1965) 1004-1006.

[7] L.P.Belluce,W.A.Kirk and E.F.Steiner, Normal structure in Banach spaces. Pacif.J.Math.26(1968) 433-440.

[8] M.A.Krasnoselskii, Positive solutions of Operator Equations. P.Noordhoff, Groningen 1964.

[9] C.A.Berger and J.G.Stampfli, Mapping Theorems for the Numerical Range. Amer.J.Math.89(1967) 1047-1055.

[10] V.I.Istrăţescu, Introduction to Fixed Point Theory. Ed.Acad. R.S.R. Bucuresti, 1973. (in Rumanian).

[11] H.Weyl, Almost periodic invariant vector sets in a metric vector space. Amer.J.Math. 71(1949) 178-205.

[12] A.Lebow and M.Schechter, Semigroups of Operators and Measures of Noncompactness. J.of Funct.Anal. 7(1971) 1-26.

[13] G.Darbo, Punti uniti in transformazioni a codominio non compatto Rend.Mat. Padova 24(1955) 84-92.

[14] T.Kato, Perturbation Theory for Linear Operators. Springer, 1966.

[15] F.Browder, On the spectral theory of eliptic differential operators. Math.Ann.142(1961) 22-130.

Fachbereich Mathematik
Der Johann Wolfgang Goethe
Uniwersität
6000 Frankfurt am Main 1
W. G e r m a n y

INTERSECTION OF THE CLASSES OF S-SELFDECOMPOSABLE AND S-SEMI-STABLE DISTRIBUTIONS

Zbigniew J. Jurek

1. Introduction. Let H be a separable real Hilbert space with the scalar product $(\cdot,\cdot)$ and the norm $\|\cdot\|$. For every non-negative r we define a shrinking operation U_r (shortly: s-operation) from H onto itself by means of the formula

$$U_r x = \begin{cases} 0 , & \text{if } \|x\| \leq r ; \\ (1 - \frac{r}{\|x\|})\, x , & \text{if } \|x\| > r . \end{cases} \tag{1.1}$$

Of course, the family U_t $(t \geq 0)$ forms a semi-group under the composition and $U_t\, U_s = U_{t+s}$ $(t,s \geq 0)$.

Let $X_1, X_2, \ldots$ be a sequence of independent H-valued random variables which are not essentially bounded in common. Further, let $r_1, r_2, \ldots$ be a non-desreasing sequence of positive numbers such that the random variables $U_{r_n} X_j$ $(k=1,2,\ldots,; n=1,2,\ldots)$ form a uniformly infinitesimal triangular array and let $a_1, a_2, \ldots$ be a sequence of vectors in H. Limit distributions of

$$Y_n = \sum_{j=1}^{n} U_{r_n} X_j + a_n \tag{1.2}$$

will be called s-selfdecomposable distributions. If we assume that the random variables $X_1, X_2, \ldots$ are identically distributed and a limit distribution of $Y_1, Y_2, \ldots$ exists for some increasing subsequence $k_1, k_2, \ldots$ of positive integers such that $k_{n+1}/k_n \to q$ and $q < \infty$, then this limit distribution will be called s-semi-stable distribution.

It follows from the definition that s-selfdecomposable distributions are infinitely divisible. Consequently, their characteristic functions are of the form

$$\exp\{i(y,x_0) - \tfrac{1}{2}(Dy,y) + \int_{H\setminus\{0\}} [e^{i(y,x)} - 1 - \frac{i(y,x)}{1+\|x\|^2}]\, M(dx)\} \tag{1.3}$$

where x_0 is a fixed element of H, D is an S-operator, M is a σ-finite measure with finite mass outside every neighborhood of the origin and

$$\int_{\|x\|\leq 1} \|x\|^2 M(dx) < \infty$$

Moreover. this representation is unique ([4],Chapter VI, Theorem 4.10). In the sequel infinitely divisible distribution with characteristic function of the form (1.3) will be denoted by $[x_0,D,M]$.

For any finite measure m on H the compound Poisson distribution e(m) is given by the formula

$$e(m) = e^{-m(H)} \sum_{k=0}^{\infty} \frac{m^{*k}}{k!}$$

where the power is taken in the sense of convolution, $m^{*0} = \delta_0$ and δ_x $(x \in H)$ denotes the probability measure concentrated at the point x.

If μ is a probability measure then by $\hat{\mu}$ we denote its characteristic function

$$\hat{\mu}(y) = \int_H e^{i(y,x)} \mu(dx) ; \qquad y \in H.$$

For any $t \geq 0$ and a measure λ let $U_t\lambda$ denote the measure defined by the formula

(1.4) $$(U_t\lambda)(E) = \lambda(U_t^{-1}E)$$

for all Borel subset E of H. It is easy to check the equations

$$U_t(U_s\lambda) = (U_tU_s)\lambda = U_{t+s}\lambda$$

for all non-negative t,s and all measures λ.

In [3] the following propositions have been proved:

<u>Proposition 1.1</u>. A probability measure μ on H is s-self-decomposable distribution if and only if $\mu = [x_0,D,M]$ and for every $t \geq 0$ and every Borel subset E of $H \setminus \{0\}$ the inequality

$$M(E) \geq (U_tM)(E)$$

is fulfilled.

<u>Proposition 1.2</u>. A probability measure μ on H is s-semi-stable distribution if and only if either $\mu = [x_0,D,0]$ (i.e. μ is a Gaussian measure) or $\mu = e(M)*\delta_{x_0}$ (i.e. μ is a shifted compound

Poisson measure) and there exist $0 < \tau < 1$ and $0 < d < \infty$ such that for all Borel subsets E of $H\setminus\{0\}$ the equation

$$(U_d M)(E) = \tau M(E)$$

holds.

The aim of this paper is to describe the intersection of the classes of s-selfdecomposable distributions and s-semi-stable distributions in terms of characteristic functions. The proof of the main theorem is based on the extreme point method (on Choquet's theorem, see [5]), adapted to the probability theory by D.G.Kendall, S.Johansen and K.Urbanik (we refer the reader to K.Urbanik [6]). Further, the Remarks 2.1 and 2.2 and the Lemma 2.4. in Section 2 are suggested by some recent work of E.Hensz and R.Jajte [2].

2. An extreme point method. Let R^+ denote the set of all positive numbers and $\bar{R}^+$ be the compactified positive real line: $\bar{R}^+ = R^+ \cup \{0\} \cup \{+\infty\}$. Further, let B and S denote the closed unit ball and unit sphere of H. If B is endowed with the relative weak-*topology of H, then the space $Q = B \times \overline{R^+}$ becomes a compact metric space. The elements of Q will be denoted by $< x,t >$ and let us put

(2.1) $$|< x,t >| = t.$$

We define a one-parameter semi-group V_t $(t \geq 0)$ of transformations of Q by assuming

(2.2) $$V_t(< x,s >) = < x, \max(0,s-t) >,$$

and transformation π from $H\setminus\{0\}$ into Q by formula

(2.3) $$\pi(x) = < \frac{x}{\|x\|}, \|x\| >$$

If B were given the relative norm topology of H, then π would be a homeomorphism of $H\setminus\{0\}$ onto $S \times R^+$. Thus, since it is well known that the Borel fields of H with respect to the norm topology and with respect to the weak-*topology concide, π and its inverse on $S \times R^+ \subset Q$ are measurable. Moreover

(2.4) $$|\pi(x)| = \|x\|,$$

and for all $x \in H\setminus\{0\}$ and all $t \geq 0$,

(2.5) $$\pi(U_t^{-1}x) = V_t^{-1}\pi(x).$$

Further, if $x \in H\setminus\{0\}$ and $0 \leq t < \|x\|$ then

(2.6) $$\pi(U_t x) = V_t \pi(x).$$

Let $\mathcal{M}_{d,\tau}$ be the set of all finite Borel measures λ on Q satisfying the conditions:

(2.7) $$\lambda(E) - V_t\lambda(E) \geq 0$$

for all $t \geq 0$, and

(2.8) $$V_d\lambda(E) = \tau\lambda(E) .$$

where $0 < d < \infty$, $0 < \tau \leq 1$ and E is an arbitrary Borel subset of $Q \setminus \{z \in Q: |z| = 0\}$. It is clear that the set $\mathcal{M}_{d,\tau}$ is convex. Let $\mathcal{K}_{d,\tau}$ be the subset of $\mathcal{M}_{d,\tau}$ consisting of probability measures. Then we have

Lemma 2.1. The set $\mathcal{K}_{d,\tau}$ is convex and compact.

Proof. It suffices to prove that $\mathcal{K}_{d,\tau}$ is closed (see [4],Chapter II, Theorem 6.4). Let $\lambda_n \in \mathcal{K}_{d,\tau}$ and $\lambda_n \Rightarrow \lambda$. Of course, the conditions (2.7) and (2.8) fulfilled for continuity sets E of λ . These sets form a field ([1],p.15) and thus (2.7) and (2.8) holds for all Borel subset E of the set $\{<x,t> \in Q: t > 0\}$. Thus, the lemma is proved.

By standard reasons (e.g.[5],[3]) we have the following lemma

Lemma 2.2. The extreme points of $\mathcal{K}_{d,\tau}$ are measures concentrated on one of the following sets: $\{<x,0>\}$, $\{<x,\infty>\}$, $\{<x,t>: t \in R^+\}$ where $x \in B$.

We proceed now to investigation of extreme points of $\mathcal{K}_{d,\tau}$ concentrated on the set $F_x = \{<x,t>: t \in R^+\}$. Let λ be a probability measure concentrated on F_x. Put

(2.9) $$J_\lambda(u) = \lambda(\{<x,t>: t \geq u\}) \qquad (u \in R^+)$$

Of course, $\lambda \in \mathcal{K}_{d,\tau}$ if and only if the conditions (2.7) and (2.8) hold for all subsets E of the form $\{<x,t>: a \leq t < b\}$, where $0 < a < b$. Taking into account (2.2) we have that $\lambda \in \mathcal{K}_{d,\tau}$ if and only if for every triplet a,b,t of positive numbers satisfying conditions $a < b$, the inequality

(2,10) $$J_\lambda(a) - J_\lambda(b) - J_\lambda(a+t) + J_\lambda(b+t) \geq 0$$

and the equality

(2.11) $$J_\lambda(a+d) = \tau J_\lambda(a)$$

are fulfilled. Substituting $b = a+t$ into (2.10) we get the inequality

$$J_\lambda(a+t) \leq \frac{1}{2}\left(J_\lambda(a) + J_\lambda(a+2t)\right)$$

for all positive a and t. Thus the function J_λ is convex. Moreover, by (2.9) it is also monotone non-increasing with $J_\lambda(+\infty) = 0$. Consequently

$$J_\lambda(u) = \int_u^\infty q_\lambda(t)dt \,, \qquad (u \in R^+) \,,$$

where the function q_λ is non-negative and monotone non-increasing. Further, by (2.11) we have that

(2.12) $$q_\lambda(u+d) = \tau q_\lambda(u)$$

for almost all $u \in R^+$. Thus q_λ is positive on R^+ almost surly, and $\int_0^\infty q_\lambda(t)dt = 1$, (since λ is probability measure). One can assume that q_λ is continuous from the left. Then q_λ is uniquely determined by λ and the condition (2.12) holds for all $u \in R^+$.

Conversly, if q is a positive monotone non-increasing, continuous from the left function, which satisfies the conditions

(2.13) $$q(u+d) = \tau q(u) \quad \text{for} \quad u \in R^+ \,,$$

and

(2.14) $$\int_0^\infty q(t)dt = 1 \,,$$

then the formula

$$\int_{F_x} f(z)\lambda(dz) = \int_0^\infty f(< x,t >)q(t)dt \,,$$

where f is any bounded continuous function on F_x, defines the probability measure λ such that $q_\lambda = q$. Thus we proved the following lemma

<u>Lemma 2.3</u>. There exists a one-to-one correspondence between all measures λ from $K_{d,\tau}$ concentrated on F_x and all positive monotone non-increasing, continuous from the left functions q on R^+ which satisfy (2.13) and (2.14).

The correspondence in question preserves convex combinations, and hence extreme points of $\mathcal{K}_{d,\tau}$ concentrated on F_x are transformed onto extreme functions in the class $\mathcal{F}$ of all positive, monotone non-increasing, continuous from the left functions on R^+ which satisfy the conditions (2.13) and (2.14). Hence the problem has been reduced to finding all non-decomposable functions in the class $\mathcal{F}$.

The function $q \in \mathcal{F}$, due to (2.13), is determined by its values on the interval $(0,d]$. More precisely, if q is a function defined on the interval $(0,d]$, positive, monotone, non-increasing, continuous from the left and such that

(2.15) $$\tau q(0+) \leq q(d) ,$$

and

(2.16) $$\int_{(0,d]} q(t)dt = 1 - \tau ,$$

then the formula (2.13) extends q to the function defined on R^+ and belonging to $\mathcal{F}$. For this reason and for convenience, in the sequel, we shall identify functions from $\mathcal{F}$ with their restrictions to the interval $(0,d]$.

We proceed now to investigation of all non-decomposable functions in the class $\mathcal{F}$.

Remark 2.1. If $q \in \mathcal{F}$ and q is a non-decomposable function then for each $t \in (0,d)$ a function q is constant on the interval $(0,t]$ or $(t,d]$.

Proof. In contrary, let us assume that q is a non-decomposable function in $\mathcal{F}$ and there exists $t_0 \in (0,d)$ such that q is not constant on both intervals $(0,t_0]$ and $(t_0,d]$. Then we get inequalities

(2.17) $$0 < q(d) < q(t_0+) \leq q(t_0) < q(0+) .$$

Let us put

$$p_1(t) = \begin{cases} A & \text{for } 0 < t \leq t_0 , \\ q(t) - B & \text{for } t_0 < t \leq d , \end{cases}$$

and

$$p_2(t) = \begin{cases} q(x) - A & \text{for } 0 < t \leq t_0 , \\ B & \text{for } t_0 < t \leq d , \end{cases}$$

where $A = (1-\tau)^{-1}[q(t_o) - \tau q(0+)]$ and $B = \tau(1-\tau)^{-1}[q(0+)-q(t_o)]$.
Taking into account (2.17), it is easy to see that

$$A \geqslant q(t_o+) - B > 0 \quad \text{and} \quad q(t_o) - A \geqslant B > 0 .$$

Thus the functions p_i (i=1,2) are positive, non-increasing, continuous from the left and moreover

$$\tau p_i(0+) \leqslant p_i(d) , \qquad i=1,2.$$

In fact, this is implied by the inequality

$$\tau A + B = \tau q(0+) \leqslant q(d).$$

Of course $p_1 + p_2 = q$ and for arbitrary positive η_1, η_2, $\eta_1 p_1 \neq \eta_2 p_2$.
Now, putting

$$\varepsilon = (1-\tau)^{-1} \int_0^d p_1(t)dt$$

we get $0 < \varepsilon < 1$, and taking $q_1 = \varepsilon^{-1} p_1$, $q_2 = (1-\varepsilon)^{-1} p_2$ we obtain $q = \varepsilon q_1 + (1-\varepsilon) q_2$ where $q_i \in \mathcal{F}$ (i=1,2) and $q_1 \neq q_2$. Thus q is not extreme element in $\mathcal{F}$ and Remark is proved.

By Remark 2.1. it follows that the non-decomposable functions in $\mathcal{F}$ may be only of the form

$$(2.18) \qquad q(u) = \begin{cases} \alpha & \text{for} \quad 0 < u \leqslant t_o , \\ \beta & \text{for} \quad t_o < u \leqslant d , \end{cases}$$

where $t_o \in (0,d]$ and

$$\alpha \geqslant \beta, \qquad \tau\alpha \leqslant \beta , \qquad \alpha t_o + \beta(d-t_o) = 1 - \tau .$$

The equality $t_o = d$ means that $q(u) = d^{-1}(1-\tau)$ for $u \in (0,d]$.

<u>Remark 2.2</u>. If q of the form (2.18) is non-decomposable function in $\mathcal{F}$ and $t_o \in (0,d)$ then $\tau\alpha = \beta$ or $\alpha = \beta$.

Proof. In contrary, let us assume that q is non-decomposable function in $\mathcal{F}$ and $\tau\alpha < \beta$ and $\alpha > \beta$. Let us put

$$p_1(u) = \begin{cases} \frac{1}{2}\alpha & \text{for} \quad 0 < u \leqslant t_o ; \\ \frac{1}{2}\beta - \gamma & \text{for} \quad t_o < u \leqslant d ; \end{cases} \qquad p_2(u) = \begin{cases} \frac{1}{2}\alpha & \text{for} \quad 0 < u \leqslant t_o ; \\ \frac{1}{2}\beta + \gamma & \text{for} \quad t_o < u \leqslant d , \end{cases}$$

where $\gamma > 0$ is chosen so that $\gamma < \frac{1}{2}(\alpha-\beta)$ and $\tau\alpha < \beta-2\gamma$. Then p_i $(i=1,2)$ are positive, monotone non-increasing, continuous from the left and $\tau p_i(0+) \leq p_i(d)$ for $i=1,2$. Now putting

$$\varepsilon = (1-\tau)^{-1} \int_0^d p_i(u)du$$

we get $0 < \varepsilon < 1$ because $p_1+p_2 = q$. Further, taking $q_1(u) = \varepsilon^{-1}p_1(u)$, $q_2(u) = (1-\varepsilon)^{-1}p_2(u)$ we obtain $q = \varepsilon q_1+(1-\varepsilon)q_2$ where $q_i \in \mathcal{F}$ $(i=1,2)$. To notice that $q_1 \neq q_2$ it suffices to consider the expressions $q_i(t_o)$ and $q_i(d)$ for $i=1,2$. Thus q is decomposable function in $\mathcal{F}$ and Remark 2.2 is proved.

Lemma 2.4. Non-decomposable functions in $\mathcal{F}$ coincide with the following one-parameter family of functions

$$q_t(u) = \begin{cases} \dfrac{1-\tau}{d+t(1-\tau)} & \text{for} \quad 0 < u \leq t\,, \\[2ex] \dfrac{\tau(1-\tau)}{\tau d+t(1-\tau)} & \text{for} \quad t < u \leq d\,, \end{cases} \tag{2.19}$$

where $t \in (0,d]$. The limit case $t = d$ means that q_d is constant on $(0,d]$.

Proof. By Remarks 2.1. and 2.2. a function q is non-decomposable if q is of the form (2.18) for $t_o \in (0,d)$ and either $\tau\alpha = \beta$ or $\alpha = \beta$. In case $\alpha = \beta$ we have that

$$q(u) = \frac{1-\tau}{d} \qquad \text{for} \quad 0 < u \leq d\,,$$

thus $q = q_d$. If $\tau\alpha = \beta$ and q is of the form (2.18) where $t_o \in (0,d)$ then in view of (2.16) we get that $q(u) = q_{t_o}(u)$.

Now it is enough to prove that the functions (2.19) are not decomposable functions in $\mathcal{F}$. Let $t \in (0,d)$ and assume that $q_t = cp_1+(1-c)p_2$ where $p_i \in \mathcal{F}$ $(i=1,2)$ and $0 < c < 1$. If one of the functions p_1 or p_2 were decreasing on $(0,t]$ or $(t,d]$ then the other one would have to increase which is imposible. Thus the functions p_1 and p_2 are constant on $(0,t]$ and $(t,d]$. Let

$$p_i(u) = \begin{cases} \alpha^{(i)} & \text{for} \quad 0 < u \leq t\,, \\ \beta^{(i)} & \text{for} \quad t < u \leq d\,, \end{cases}$$

where $i=1,2$. Of course we have $\alpha^{(i)} \geq \beta^{(i)}$ and

$$(2.20) \qquad \tau\alpha^{(i)} \leq \beta^{(i)} \quad , \qquad (i=1,2).$$

If any of the inequalities (2.20) were sharp then we would have

$$\tau q_t(0+) = c\tau\alpha^{(1)}+(1-c)\tau\alpha^{(2)} < c\beta^{(1)}+(1-c)\beta^2 = q_t(d),$$

which contradicts with $\tau q_t(0+) = q_t(d)$. Thus $\tau\alpha^{(i)} = \beta^{(i)}$ for $i=1,2$ and from conditions $\int_0^d p_i(u)du = 1-\tau$, $(i=1,2)$ we get $p_1 = p_2 = q_t$. Hence the functions (2.19) for $t \in (0,d)$ are non--decomposable elements in $\mathcal{F}$.

In case $t=d$ (i.e. $q_d(u) = (1-\tau)d^{-1}$) the proof is similar. Thus the Lemma is proved.

We define a family $m_{<x,t>}$ $(t \in (0,d])$ of probability measures on F_x which are determined by the functions (2.19) via Lemma 2.3. Thus for any continuous bounded function f on F_x we have the formula

$$(2.21) \qquad \int_{F_x} f(z)m_{<x,t>}dz =$$

$$= \frac{1-\tau}{\tau d+|<x,t>|(1-\tau)} \sum_{k=0}^{\infty} \tau^k \int_{(0,d]} f(V_{kd}^{-1}(<x,s>)\Psi_{|<x,t>|}(s)ds$$

where

$$(2.22) \qquad \Psi_{|<x,t>|}(s) = 1_{(0,|<x,t>|]}(s) + \tau 1_{(|<x,t>|,d]}(s)$$

and 1_A denotes the indicator of a set A.

Further we put

$$(2.23) \qquad m_{<x,\alpha>} = \delta_{<x,\alpha>} \qquad \text{if either } \alpha = 0 \text{ or } \alpha = \infty$$

and let

$$D = B \times ([0,d] \cup \{+\infty\})$$

From (2.21) and (2.23) we have that the mapping $z \rightarrow m_z$ from D onto the set of extreme points of $\mathcal{K}_{d,\tau}$ is not continuous in points of the form $<x,0>$ where $x \in B$. Further, the set $\{<x,0> : x \in B\}$ is homeomorphic to the set $\{m_{<x,o>} : x \in B\}$ (see [4],Chapter II, Lemma 6.1). Taking into account Corollary

3.3, Chapter I in [4] we have that the set $\{m_z : z \in D\}$ is a Borel subset of the space of all Borel measures on Q. Moreover the mapping $z \to m_z$ is Borel isomorphism. Thus we have the following lemma.

Lemma 2.5. The set m_z $(z \in D)$ defined by (2.21) and (2.23) coincides with the set of extreme points of $\mathcal{K}_{d,\tau}$. Moreover the mapping $z \to m_z$ is a Borel isomorphism between D and the set of extreme points of $\mathcal{K}_{d,\tau}$.

Once the extreme points of $\mathcal{K}_{d,\tau}$ are found we can apply a Theorem by Choquet (see [5],Chapter 3). Since each element of $\mathcal{K}_{d,\tau}$ is of the form cv, where $c > 0$ and $v \in \mathcal{K}_{d,\tau}$, we then get the following lemma

Lemma 2.6. A measure μ belongs to $\mathcal{M}_{d,\tau}$ if and only if there exists a finite Borel measure λ on D such that for each continuous function f on Q the equation

$$\int_Q f(u)\mu(du) = \int_D \int_Q f(u)m_z(du)\lambda(dz)$$

holds.

3. A representation of characteristic functions. By Propositions 1.1. and 1.2. (in Section 1) we have the following lemma.

Lemma 3.1. A probability measure μ on H is simultaneously s-selfdecomposable and s-semi-stable distribution on H if and only if either μ is a Gaussian measure i.e.

$$\hat{\mu}(y) = \exp\{i(y,x_0) - \frac{1}{2}(Dy,y)\}$$

where x_0 is an element of H and D is an S-operator, or μ is a shifted compound Poisson measure i.e.

$$\hat{\mu}(y) = \exp\{i(y,x_0) + \int_{H\setminus\{0\}} [e^{i(y,x)} - 1]M(dx)\}$$

where M is finite on $H\setminus\{0\}$ and for all Borel subsets E of $H\setminus\{0\}$ and for every $t \geq 0$

$$M(E) \geq (U_t M)(E)$$

and there exists $0 < \tau < 1$, $0 < d < \infty$ such that

$$(U_d M)(E) = \tau M(E) .$$

Now we prove the following statement

Lemma 3.2. A Poissonian measure $\mu = e(M) * \delta_{x_0}$ on H is simultaneously s-selfdecomposable and s-semi-stable distribution on H if and only if the the measure πM belongs to $\mathcal{M}_{d,\tau}$.

Proof. Let E be a Borel subset of $Q \setminus \{z \in Q: |z| = 0\}$. Taking into account (2.5) and Lemma 3.1 we have

$$\pi M(E) - V_t \pi M(E) = \pi M(E) - U_t \pi M(E) \geqslant 0$$

and

$$V_d \pi M(E) = U_d \pi M(E)$$

Consequently, $\mu = e(M) * \delta_{x_0}$ is simultaneously s-seldecomposable and s-semi-stable distribution if and only if $\pi M \in \mathcal{M}_{d,\tau}$.
Thus the Lemma is proved.

Theorem 3.1. A function φ on H is the characteristic function of a Poissonian measure which is simultaneously s-self-decomposable and s-semi-stable distribution on H if and only if

$$\varphi(y) = \exp\{i(y,x_0) + \int_{0<\|x\|\leqslant d} [\sum_{k=0}^{\infty} \tau^k \int_{(kd,(k+1)d]} (e^{i(y,\frac{x}{\|x\|})t} - 1)\Psi_{\|x\|}(t-kd)dt] \rho(dx)\} \tag{3.1}$$

where x_0 is an element of H; τ,d are constants such that $0 < \tau < 1$, $0 < d < \infty$; ρ is a finite Borel measure on the set $\{x \in H: 0<\|x\| \leqslant d\}$ and $\Psi_{\|x\|}$ is given by the formula

$$\Psi_{\|x\|}(s) = 1_{(0,\|x\|]}(s) + \tau 1_{(\|x\|,d]}(s) , \quad s \in (0,d]. \tag{3.2}$$

Proof. The necessity. Let a shifted compound Poisson measure $\mu = e(M) * \delta_{x_0}$ will be simultaneously s-selfdecomposable and s-semi--stable distribution on H. By Lemma 3.2. we have that πM belongs to $\mathcal{M}_{d,\tau}$. Further, by Lemma 2.6. there exists a finite Borel measure λ on $D = B \times ([0,d]\cup\{\infty\})$ such that for every continuous function f on Q the equation

$$(3.3) \qquad \int_Q f(u)\pi M(du) = \int_D\int_Q f(u)m_z(du)\lambda(dz)$$

holds. Here m_z $(z\in D)$ denotes the extreme points of $\mathcal{K}_{d,\tau}$ defined by the formulae (2.21) and (2.23). It is clear that the measure πM is concentrated on the set $S\times R^+$. Consequently, by (3.3), the measure λ is concentrated on the set $D_0 = S\times(0,d]$. Since for $z\in D_0$ the measures m_z are concentrated on $S\times R^+$ (see 2.21) then formula (3.3) can be rewritten in the form

$$(3.4) \qquad \int_{S\times R^+} f(u)\pi M(du) = \int_{D_0}\int_{S\times R^+} f(u)m_z(du)\lambda(dz)$$

for any function f continuous and bounded on $S\times R^+$. Let us introduce the notation $v = \pi^{-1}\lambda$ and finite Borel measure ρ as follows

$$\rho(E) = \int_E \frac{v(dx)}{d+(1-\tau)\|x\|}$$

where E is a Borel subset of $\{x:\ 0<\|x\|\leq d\}$. For every continuous and bounded function g on $H\setminus\{0\}$, in view of (1.1),(2.1)-(2.5), (2.21) and (3.4) we get the formula

$$\int_{H\setminus\{0\}} g(x)M(dx) = \int_{0<\|x\|\leq d}\Big[\sum_{k=0}\tau^k\int_{(0,d]} g((s+kd)\frac{x}{\|x\|})\Psi_{\|x\|}(s)ds\Big]\rho(dx)$$

where

$$\Psi_{\|x\|}(s) = 1_{(0,\|x\|]}(s) + \tau 1_{(\|x\|,d]}(s)\ , \qquad s\in(0,d].$$

Setting

$$g(x) = e^{i(y,x)} - 1$$

where $y\in H$, into the last formula, after calculation we obtain the formula (3.1) which completes the proof of the necessity.

The sufficiency. For $x\in H\setminus\{0\}$ and non-negative integer k we define the measures $m_x^{(k)}$ on H by the formula

$$(3.5) \qquad m_x^{(k)}(E) = \Big|\{s:\ s\frac{x}{\|x\|}\in E\wedge kd<s\leq(k+1)d\}\Big|$$

where E is a Borel subset of H and $|\cdot|$ denotes the Lebesgue measure on positive half-line. Further, for $x \in H$ such that $0 < \|x\| \leq d$ let

$$(3.6) \qquad \varphi_{\|x\|}(\|z\|) = \begin{cases} 1 & \text{if } kd < \|z\| \leq kd + \|x\| , \\ \tau & \text{if } kd + \|x\| < \|z\| \leq (k+1)d , \end{cases}$$

where $z \in H$ and let

$$(3.7) \qquad m_x = \sum_{k=0}^{\infty} \tau^k m_x^{(k)} .$$

Of course m_x is finite Borel measure concentrated on the set $\{t \frac{x}{\|x\|} : t \in R^+\}$. Now we define

$$(3.8) \qquad M_x(E) = \int_E \varphi_{\|x\|}(\|z\|) m_x(dz)$$

and

$$(3.9) \qquad M(E) = \int_{0<\|x\|\leq d} M_x(E)\rho(dx)$$

for every Borel subset E of H. By a simple computation, in view of (3.5)-(3.9) we obtain that

$$\varphi(y) = \exp[i(y,x_0) + \int_H (e^{i(y,z)} - 1)M(dz)]$$

and M is finite Borel measure on H. Thus φ is the characteristic function of the measure $\mu = e(M) * \delta_{x_0}$.

It remains to prove that μ is simultaneously s-semi-stable and s-selfdecomposable distribution. Let E be a Borel subset of $H \setminus \{0\}$. Taking into account (3.5) and (3.7) we get

$$U_d m_x^{(k)}(E) = |\{t: (k-1)d < t-d \leq kd \wedge (t-d)\tfrac{x}{\|x\|} \in E\}| = m_x^{(k-1)}(E)$$

for $k \geq 1$, and

$$(3.10) \qquad U_d m_x(E) = \sum_{k=1}^{\infty} \tau^k m_x^{(k-1)}(E) = \tau m_x(E).$$

By (3.10) and (3.6) we have

$$U_d M_x(E) = \int_E \varphi_{\|x\|}(\|z\|+d) U_d m_x(dz) = \tau M_x(E)$$

Hence and from (3.9) we obtain

$$U_d M(E) = \tau M(E)$$

where E is a Borel subset of $H\setminus\{0\}$. Thus μ is s-semi-stable distribution (see Proposition 1.2).

Let $0 < s < d$ be fixed and E be a Borel subset of $\{t \frac{x}{\|x\|} : kd < t \leq (k+1)d\}$. Then $U_s^{-1}E \subset \{t \frac{x}{\|x\|} : kd+s < t \leq (k+1)d+s\}$ and

$$U_s m_x(E) = \tau^k m_x^{(k)}(U_s^{-1}E) + \tau^{k+1} m_x^{(k+1)}(U_s^{-1}E) \leq$$

$$\leq \tau^k |\{t: t \frac{x}{\|x\|} \ U_s^{-1}E \wedge kd+s < t \leq (k+1)d+s\}| = \tau^k m_x^{(k)}(E) = m_x(E).$$

If E is an arbitrary Borel subset of $\{t \frac{x}{\|x\|}: t \in R^+\}$ then $U_s^{-1}E \subset \{t \frac{x}{\|x\|}: t > s\}$ and

$$m_x(E) = \sum_{k=0}^{\infty} m_x(E \cap \{t \frac{x}{\|x\|} : kd < t \leq (k+1)d\}) \geq$$

$$\geq \sum_{k=0}^{\infty} m_x(U_s^{-1}E \cap \{t \frac{x}{\|x\|} : kd+s < t \leq (k+1)d+s\}) = U_s m_x(E).$$

Hence and from (3.10) for every $t > 0$ and every Borel subset E of $H\setminus\{0\}$ we have

$$U_t m_x(E) = \tau^k U_s m_x(E) \leq m_x(E) \tag{3.11}$$

whenever $t = kd+s$ and $0 < s < d$.

Let a Borel subset E of $\{t \frac{x}{\|x\|} : kd < t \leq (k+1)d\}$ and number s such that $0 < s < d$, be fixed. Let us define

$$E_1 = \{z \in E: \varphi_{\|x\|}(\|z\|) \geq \varphi_{\|x\|}(\|z\|+s)\}$$

Then

$$E\setminus E_1 = \{z \in E: \varphi_{\|x\|}(\|z\|) = \tau \quad \text{and} \quad \varphi_{\|x\|}(\|z\|)+s) = 1\}$$

and

$$\int_{E\setminus E_1} \varphi_{\|x\|}(\|z\|+s)U_s m_x(dz) = U_s m_x(E\setminus E_1) =$$

$$= \tau^{k+1}|\{t: t\frac{x}{\|x\|}\in E\setminus E_1\}| = \tau[\tau^k m_x^{(k)}(E\setminus E_1)] = \int_{E\setminus E_1} \varphi_{\|x\|}(\|z\|)m_x(dz).$$

From the last equality, (3.8) and (3.11) we get

$$U_s M_x(E) \leqslant M_x(E)$$

for all $0 < s < d$ and all Borel subsets E of $\{t\frac{x}{\|x\|}: kd < t \leqslant (k+1)d\}$. Quite similar as before (for the measure m_x) one can prove that the last inequality holds for all positive s and all Borel subsets E of $H\setminus\{0\}$. Hence we obtain inequality

$$M(E) \geqslant U_s M(E)$$

for $s \geqslant 0$ and Borel subset E of $H\setminus\{0\}$. Thus μ is also s-self-decomposable distribution (see Proposition 1.1), which completes the proof of Theorem.

After some simple computation we obtain from Theorem 3.1 the following characterization:

Corollary 3.1. A function φ on H is the characteristic function of a Poissonian measure which is simultaneously s-self-decomposable and s-semi-stable distribution on H if and only if

$$(3.12) \qquad \varphi(y) = \exp\{i(y,x_0) +$$

$$+ \int_{0<\|x\|\leqslant d} [(\frac{(1-\tau)e^{i(y,x)}}{1-\tau e^{i(y,\frac{x}{\|x\|})d}} - 1)\frac{1}{i(y,\frac{x}{\|x\|})} - \frac{\tau d+(1-\tau)\|x\|}{1-\tau}]\rho(dx)\}$$

where x_0 is a fixed element of H, τ, d are constant such that $0 < \tau < 1$, $0 < d < \infty$ and ρ is finite Borel measure on the set $\{x\in H: 0 < \|x\| \leqslant d\}$

As a consequence of Propositions 1.1, 1.2, Corollary 3.1. and Theorem 4.9 in [4] Chapter VI we get the following statement

Theorem 3.2. A function φ on H is the characteristic function of a probability measure which is simultaneously s-self-decomposable and s-semi-stable distribution if and only if either

$$\varphi(y) = \exp\{i(y,x_0) - \frac{1}{2}(Dy,y)\}$$

where x_0 H and D is an S-operator or φ is of the form (3.12).

4. Example. There exists s-semi-stable distribution which is not s-selfdecomposable.

Let $\mu = e(M)*\delta_{x_0}$ and M be a purely atomic measure concentrated on the points of the form

$$x_k = 2ke_0 \qquad (k=1,2,\dots)$$

where e_0 is a fixed vector from the unit sphere of H, and

$$M(\{x_k\}) = a^k \qquad (0 < a < 1)$$

It is easy to verify that $U_2M(E) = aM(E)$ for all Borel subsets E of $H\setminus\{0\}$. Thus μ is s-semi-stable distribution with d=2 and $\tau=a$, (see Proposition 1.2).

On the other hand let E be a Borel subset of $H\setminus\{0\}$ such that x_k (k=1,2,...) does not belongs to E, but for some $t>0$

$$(U_t^{-1}E) \cap \{x_k: k=1,2,\dots\} \neq \emptyset$$

Then $M(E)-U_tM(E)< 0$ and in view of Proposition 1.1 we have that the probability measure μ is not s-selfdecomposable distribution, Thus the Example is proved.

Of course an arbitrary s-selfdecomposable distribution with simultaneously non-vanishing Gaussian and Poissonian components is not s-semi-stable distribution.

References

[1] P.Bilingsley, Convergence of probability measures, New York 1968.

[2] E.Hensz,R.Jajte, On a class of limit laws, Theory of Probability and its Applications, (in print).

[3] Z.Jurek, Limit distributions for sums of shrunken random variables, Dissertationes Mathematicae (Rozprawy matematyczne), in preparation.

[4] K.R.Parthasarathy, Probability measures on metric spaces, New York - London 1967.

[5] R.R.Phelps, Lectures on Choquet's theorem, Princeton 1966.

[6] K.Urbanik, Extreme point method in probability theory, Lecture Notes in Mathematics No 472 (1975), 169-194.

Wrocław University
Institute of Mathematics
Pl.Grunwaldzki 2/4
50-384 Wrocław, P o l a n d

PROPAGATORS AND DILATIONS

P. Masani

1. Introduction.

Recent work has revealed a close nexus between the concepts of propagator and dilation, and the importance in this connection of the operator-valued Kernel Theorem culminating from the work of Kolmogorov [13], Aronszajn [1,2] (especially) and Pedrick [23]. A unified theory embracing both concepts is presented in our recent paper [18]. In the present paper we attempt to organize the new insights which have been gained in this field since the completion of [18].

We find that of the two concepts, propagator and dilation, it is the former which is more fundamental : all dilation theorems are easily deduced via the Kernel Thm. from those on propagators, once the hard step of finding the appropriate positive-definite kernel is accomplished. Consequently, the bulk of this paper, §§ 2-5, deals with Hilbertian varieties X(.), their covariance kernels K(. , .), and their propagators S(.). Our objective is to obtain the necessary and sufficient conditions that K(. , .) must fulfill in order that S(.) may exist. Most of this work has already been done in [18]. Here our primary effort is directed to demonstrating the equivalence of the conditions found in [18] to those obtained independently by F.H. Szafraniec [27]. In § 6 we state several dilation theorems which are deducible from ones on propagators by appeal to the Kernel Theorem, and in 6.7 we lay down a routine whereby this can be done in general.

Finally in § 7 we refer to some of the bibliography on the subject, which bears on these developments, and some unpublished work of Shonkwiler and Faulkner [25] in which they attempt to widen the theory by the incorporation of possibly discontinuous linear operator-valued PD kernels K(. , .) on $\Lambda x \Lambda$ where Λ is less than an involutory semi-group.

We conclude this section by explaining the notation we will employ and stating a few ancillary concepts and results.

$\mathbb{F}$ will refer to etheir the real number field $\mathbb{R}$ or the complex number field $\mathbb{C}$, and $\mathbb{N}$ to the set of all integers. $\mathbb{N}_+$, $\mathbb{R}_+$, and $\mathbb{N}_{0+}$, $\mathbb{R}_{0+}$ will denote the subsets of positive elements, and subsets of non-negative elements of $\mathbb{N}$ and $\mathbb{R}$.

Let $\mathfrak{X}$, Y be normed vector spaces over $\mathbb{F}$. Then $\forall A \subset \mathfrak{X}$, $\langle A \rangle$ and $\sigma(A)$ will denote the linear manifold spanned by A, and the closed linear manifold spanned by A, respectively. $CL(\mathfrak{X}, Y)$ will denote the class of continuous linear operators on $\mathfrak{X}$ to Y. $\mathfrak{X}^*$ will denote the adjoint of $\mathfrak{X}$, i.e.

$$\mathfrak{X}^* = \{\overline{f(.)} : f(.) \in CL(\mathfrak{X}, \mathbb{F})\}.$$

Thus $\mathfrak{X}^*$ is the dual of $\mathfrak{X}$, iff. $\mathbb{F} = \mathbb{R}$. For a Hilbert space we shall identify $\mathcal{H}$ and $\mathcal{H}^*$.

For an operator $A \in CL(\mathfrak{X}, Y)$, the adjoint A^* of A is defined to be the Y^*-to $\mathfrak{X}^*$ operator such that

$$A^*(f) = f \circ A, \quad f \in Y^* .$$

It follows that if $J \in CL(W, \mathcal{H})$, then $J^*J \in CL(W, W^*)$ and

$$\forall \; w_1, w_2 \in W, \quad [(J_2^* J_1)(w_1)](w_2) = (J_1 w_1, J_2 w_2)_{\mathcal{H}} .$$

Definition 1.1 . Let W be a Banach space over $\mathbb{F}$.

(a) An operator $H \in CL(W, W^*)$ is called hermitian, iff. $\forall w, w' \in W$, $\{H(w')\}(w) = \overline{\{H(w)\}(w')}$.

(b) An operator $H \in CL(W, W^*)$ is called non-negative (in symbols $H \geq 0$), iff. H is hermitian & $\forall w \in W$, $\{H(w)\}(w) \geq 0$.

(c) For $H_1, H_2 \in CL(W, W^*)$ we write $H_1 \geq H_2$ to mean that $H_1 - H_2 \geq 0$.

We refer to [18: 2.4] for the simple properties of hermitian operators and non-negative operators $H \in CL(W, W^*)$. An obvious relation we will need, which is not in [18], is that if $H_1, H_2 \in CL(W, W^*)$, then

$$(1.2) \qquad H_1 \geq H_2, \quad \text{iff} \; \forall w \in W, \quad [H_1(w)](w) \geq [H_2(w)](w).$$

For a topological space (Γ, τ) we shall write $Bl(\Gamma)$ for the σ-algebra generated by the topology τ, i.e. for the family of Borel subsets of Γ.

2. Hilbertian varieties and their covariance kernels

The ingredients of the theory under consideration are as follows:

$$
(2.1)\begin{cases}
\text{(i)} & \Lambda \text{ is an arbitrary set} \\
\text{(ii)} & W \text{ is a Banach space over } \mathbb{F} \\
\text{(iii)} & \quad \text{is a Hilbert space over } \mathbb{F} \\
\text{(iv)} & \Gamma \text{ is an (additive but not necessarily abelian)} \\
 & \text{semi-group (s.g.) with neutral element } 0.
\end{cases}
$$

2.2. Definition (a) By a Hilbertian variety we shall mean a function $X(\cdot)$ on Λ to $\mathcal{H}$ or a function $X(\cdot)$ on Λ to $CL(W,\mathcal{H})$ where $\mathcal{H}$ is any Hilbert space.

(b) The subspace of a Hilbertian variety $x(\cdot)$ or $X(\cdot)$ is defined by

$$\mathfrak{S}_x \underset{d}{=} \{x(\lambda) : \lambda \quad \Lambda\} \qquad \mathfrak{S}_X \underset{d}{=} \{X(\lambda)(W) : \lambda \in \Lambda\}$$

(c) By the covariance kernel of a Hilbertian variety $x(\cdot)$ or $X(\cdot)$ we shall mean the function $k(\cdot,\cdot)$ or $K(\cdot,\cdot)$ on $\Lambda \times \Lambda$ defined by

$$k(\lambda, \lambda') \underset{d}{=} (x(\lambda), x(\lambda'))_{\mathcal{H}}, \quad K(\lambda, \lambda') \underset{d}{=} X(\lambda')^{*}.X(\lambda).$$

Thus $k(\cdot\,,\cdot)$ takes values in $\mathbb{F}$ and $K(\cdot\,,\cdot)$ takes values in $CL(W,W^{*})$.

Any two Hilbertian varieties having the same covariance kernel are unitarily equivalent as the following proposition asserts [18 : 2.9] :

2.3 Congruence Thm. Let (i) W be a Banach space and $\mathcal{H},\mathcal{K}$ be Hilbert spaces over $\mathbb{F}$, (ii) $X(\cdot)$, $Y(\cdot)$ be functions on Λ to $CL(W,\mathcal{H})$, $CL(W,\mathcal{K})$, respectively, having the same covariance kernel, i.e. such that

$$\forall\, \lambda', \lambda \in \Lambda, \quad X(\lambda')^{*}X(\lambda) = Y(\lambda')^{*}Y(\lambda).$$

Then $\exists$ a unitary operator V on $\mathfrak{S}_X$ onto $\mathfrak{S}_Y$ such that $Y(\cdot) = V \circ X(\cdot)$.

Another basic fact is the close relationship between the concepts of covariance kernel and positive definite kernel in the following sense :

2.4. **Definition.** A kernel $K(\cdot\,,\cdot)$ on $\Lambda x \Lambda$ to $CL(W,W^*)$ is called positive definite (PD), iff. $\forall$ functions $C(\cdot)$ on Λ to $CL(W,W)$, $\forall\, r \in \mathbb{N}_+$, & $\forall \lambda_1,\ldots,\lambda_r \in \Lambda$,

$$\sum_{i=1}^{r} \sum_{j=1}^{r} C(\lambda_j)^* K(\lambda_i,\lambda_j) C(\lambda_i) \geq 0,$$

and ≠≠

$$\forall\, \lambda,\lambda' \in \Lambda \ \&\ \forall\, w,w' \in W,\ [K(\lambda,\lambda')(w)](w') = \overline{[K(\lambda',\lambda)(w')](w)}.$$

The trivial part of the relationship in question is that

(2.5) For any Hilbertian variety $X(\cdot)$, the covariance kernel $K_X(\cdot\,,\cdot)$ is a PD kernel on $\Lambda x \Lambda$ to $CL(W,W^*)$.

The non-trivial part is that every PD kernel on $\Lambda x \Lambda$ to $CL(W,W^*)$ is the covariance kernel of a Hilbertian variety $X(\cdot)$. This constitutes the Kernel Thm. alluded to in § 1. It plays a crucial role in dilation theory but hardly any in propagator theory, and its discussion is therefore deferred to § 6.

3. Propagators

Let $X(\cdot)$ be a Hilbertian variety on Λ. The concepts of its propagator and of its being stationary become meaningful when a semi-group Γ acts on Λ, cf.(2.1((iv). The basic concepts now concerning us are as follows :

3.1. **Definition.** Let (i) $X(\cdot)$ be a Hilbertian variety, i.e. a function on Λ to $CL(W,\mathcal{H})$, where Λ, W, $\mathcal{H}$ are as in (2.1)(i-)-(iii); (ii) the s.g. Γ be as in (2.1)(iv). Then

(a) we say that Γ acts on Λ, iff. $\exists$ a binary operation $\circledast$ on $\Gamma x \Lambda$ to Λ such that

$$\forall\, s,t \in \Gamma \ \&\ \forall \lambda \in \Lambda, \quad (s+t)\circledast\lambda = s\circledast(t\circledast\lambda) \ \&\ 0\circledast\lambda = \lambda.$$

(b) $S(\cdot)$ is called a propagator (or controller) of $X(\cdot)$, iff. $S(\cdot)$ is a function on Γ to $CL(\mathfrak{S}_X,\mathfrak{S}_X)$, where $\mathfrak{S}_X$ is as in 2.2(b), and

---

≠≠ When $\mathbb{F} = \mathbb{C}$ the condition to come follows from the last, and is therefore redundant. We leave this to the reader to check. When W is a Hilbert space, thishas the equivalent simpler rendering: $K(\lambda,\lambda') = K(\lambda',\lambda)^*$.

$$\forall t \in \Gamma \quad \forall \lambda, \quad S(t)\circ X(\lambda) = X(t\oplus\lambda).$$

(c) $X(.)$ is called <u>stationary</u>, iff. its covariance kernel $K(. , .)$ on $\Lambda\times\Lambda$, cf. 2.2(c), is translation-invariant in the sense that

$$\forall t \in \Gamma \ \& \ \forall \lambda, \lambda', \quad K(t\oplus\lambda, t\oplus\lambda') = K(\lambda,\lambda').$$

$$(3.2) \quad \begin{cases} \text{if the variety } X(.) \text{ has a propagator } S(.), \text{ then} \\ (S(t): \ t\in\Gamma) \text{ is a s.g. of operators in } CL(\mathfrak{S}_X, \mathfrak{S}_X) \text{ and} \\ S(0) = I_{\mathfrak{S}_X}. \end{cases}$$

3.3. <u>Examples.</u> The following illustrate the notion of " Γ acting on Λ" :

(i) Consider a solid body spinning about an axis with constant angular velocity. Let $\Lambda(\subseteq \mathbb{R}^3)$ be the set of position vectors of its particles, let $\Gamma = \mathbb{R}_{0+}$, and let $\forall t \in \Gamma \ \& \ \forall \lambda \in \Lambda$, $t\oplus\lambda$ be the position vector of a particle t seconds after its position vectors is λ.

(ii) Let Γ be a topological group, $\Lambda \underset{d}{=} B\,l(\Gamma)$ be the family of Borel subsets of Γ , and

$$\forall t \in \Gamma \ \& \ \forall B \in \Lambda, \quad t\oplus B \underset{d}{=} \{t + \lambda : \lambda \in B\} \in \Lambda .$$

It is easily seen that the $\oplus$ defined in both (i) and (ii) satisfies the condition in 3.1(a).

3.4. <u>Determinism, time-invariance, stationarity</u>. To understand the scope and significance of the concepts introduced in Def.3.1, look upon Λ as the phase space of a system evolving with respect to "multidimensional time" $t \in \Gamma$, i.e. look upon the s.g. Γ as a "time-domain", and regard the variety $X(.)$ as a vectorial or operational quantity associated with the system and dependent on its phase. What are the restraints imposed on the system by the requirement that Γ acts on Λ and by the one that $X(.)$ possesses a propagator $S(.)$?

To answer these questions, consider the familiar case where $\Gamma = \mathbb{R}_{0+}$ and $t \in \Gamma$ represents ordinary time. Assume that the system is deterministic, i.e. that its phase λ at instant t is determinable from its phase at any ealier instant s. Then there exists a (single-valued) transition operator $T(s,t)$, $0 \leqslant s \leqslant t$, on Λ to Λ such that

(1) $T(s,t)(\lambda)$ = the phase at instant t of the system whose phase at instant s was λ.

It follows easily that

(2) for $0 \leq s \leq t \leq u$, $T(s,u) = T(t,u).T(s,t)$,

this equality being the mathematical expression of the socalled <u>major premise of Huygen's Principle</u>, cf. [11: p. 618] and [8 : pp. 53-54].

Now assume that the deterministic system is <u>time-invariant</u>, i.e. that the transition operator $T(\cdot\ ,\ \cdot)$ is translation-invariant:

$$T(s+h,\ t+h) = T(s,t),\quad 0 < s < t \ \&\ h \in \mathbb{R}_+,$$

or equivalently that $T(s,t)$ depends only on t-s. Writing $T_{t-s} \underset{d}{=} T(s,t)$, the equation (1) and (2) give way to

(1') $T_t(\lambda)$ = the phase of the system t seconds after it is in phase λ

and

(2') $T_{s+t} = T_s \cdot T_t$, $s,t \in \mathbb{R}_{0+}$

Now write, $t \circledast \lambda$ instead of $T_t(\lambda)$. Then (1') becomes

(3) $t \circledast \lambda$ = the phase of the system t seconds after it is in the phase λ,

and from (2') we immediately get the equations in 3.1 (a); i.e. we find that $\mathbb{R}_{0+}$ acts on Λ in the sense of Def. 3.1(a). It is easy to see that conversely if the s.g. $\mathbb{R}_{0+}$ acts on Λ, then with $t \circledast \lambda$ interpreted as in (3) we have a determinictic time-invariant system. In short, when we interpret the s.g. Γ as a "time domain", <u>the condition that Γ acts on Λ amounts to dealing with a deterministic time-invariant system.</u>

Now let $\mathfrak{D} \underset{d}{=} \bigcup_{\lambda \in \Lambda} X(\lambda)(W) \subset \mathcal{H}$, and define $\forall t \in \mathbb{R}_{0+}$, the $\mathfrak{D}$-to-$\mathfrak{D}$ operator S_t by

(4) $S_t\{X(\lambda)(w) \underset{d}{=} [X\{T_t(\lambda)\}](w) \underset{d}{=} X(t \circledast \lambda)(w)$, $\lambda \in \Lambda$, $w \in W$.

Then from (2') we easily see that

(5) $S_{s+t} = S_s \cdot S_t$, $s,t \in \mathbb{R}_{0+}$.

Observe, however that the definition (4) does not entail that the linear extensions $S(t) \underset{d}{=} \langle S_t \rangle$ of the operators S_t are either single-valued or continuous on $\langle \mathfrak{D} \rangle$ to $\langle \mathfrak{D} \rangle$. The require that the variety $X(.)$ possess a propagator is ineffect to demand that $S(t) \underset{d}{=} \langle S_t \rangle$ be single-valued and continuous on $\langle \mathfrak{D} \rangle$

To see what this requirement entails, let $\mathfrak{M}$ be the class of all $\mathbb{F}$-valued, finitely additive, measures μ on $2^{\mathfrak{D}}$ having finite carriers, and $m(\mu)$ be the moment of μ about O, i.e.

$$m(\mu) \underset{d}{=} \int_{\mathfrak{D}} x\mu(dx).$$

Then the necessary and sufficient condition that each $S(t)$ be single-valued may be expressed in the form

$$(6) \qquad \forall \mu \in \mathfrak{M}, \quad m(\mu) = 0 \Rightarrow \forall t \in \mathbb{R}_{0+}, \quad m(\mu . S_t^{-1}) = 0$$

The further condition sufficient to ensure that each $S(t)$ be continuous on $\mathfrak{D}$ is that there exist a function $\beta(\cdot)$ on R_{0+} to R_{0+} such that

$$(7) \qquad \forall \mu \in \mathfrak{M} \;\&\; \forall t \in R_{0+}, \; |m(\mu \cdot S_t^{-1})| \leq \beta(t) \cdot |m(\mu)|.$$

The inequality (7) of course implies (6) and therefore the single-valuedness of each $S(t)$. Since by (4) S_t depends on $X(\cdot)$ and on $\oplus$, the inequality (7) places in effect a restraint on the operation $\oplus$ and on the variety $X(.)$. But from (7) it does not follow that the variety $X(\cdot)$ is stationary in the sense of Def.3.1(c).

To sum up, with the temporal interpretation of the s.g.Γ, the imposition that Γ act on Λ amounts to our dealing with a deterministic, time-invariant system with phase space Λ, and the imposition that the variety $X(\cdot)$ possesses a propagator amounts to placing the restraint (7), but does not necessarily make $X(\cdot)$ stationary in the sense of Def.3.1(c). Briefly, we are in a deterministic, time-invariant but not necessarily stationary situation.

We are now ready to address the central question concerning us:

3.5. Question. Let the s.g. Γ act on Λ, and let $X(\cdot)$ be a Hilbertian variety on Λ. What conditions on the covariance kernel $K_X(\cdot,\cdot)$ of $X(\cdot)$ are necessary and sufficient to ensure that $X(\cdot)$ has a propagator $S(\cdot)$?

The answer, which is nearly apparent from (7), is provided by the following theorem [18: 3.4].

3.6. Theorem. Let the s.g. Γ act on Λ . Then the variety $X(\cdot)$ on Λ has a propagator $S(\cdot)$, iff. $\exists$ a function $\lambda(\cdot)$ on Γ to $\mathbb{R}_{0+}$ such that $\forall w(\cdot) \in W^{\Lambda}$, $\forall r \in \mathbb{N}_+, \forall \lambda_1, \ldots, \lambda_r \in \Lambda$,

$$0 \leq \sum_{i=1}^{r} \sum_{j=1}^{r} [K_X(t\oplus\lambda_i, t\oplus\lambda_j)w(\lambda_i)]w(\lambda_j)$$
$$\leq \gamma(t) \sum_{i=1}^{r} \sum_{j=1}^{r} [K_X(\lambda_i, \lambda_j)w(\lambda_i)]w(\lambda_j).$$

We shall refer to the last inequality as the Nagy translational inequality, as it is a generalization of the one obtained by Nagy [21,(c)] in the special case $\Lambda = \Gamma =$ an involutory s.g., and $W =$ a Hilbert space, in order to secure for the kernel $K_X(\cdot\ ,\cdot)$ a *-representation in a superspace $\mathcal{K} \supseteq W$. It is also an extension of an inequality obtained by Getoor [7: 2.1, 2.2] in the special case $\Lambda = \Gamma = \mathbb{R}_{0+}$ and $W = \mathbb{F}$.

It follows easily from Thm. 3.6., cf. [18 : 3.5], that

(3.7) the variety $X(\cdot)$ is stationary, iff $X(\cdot)$ has a propagator such that each $S(t)$ is an isometry on $\mathcal{S}_X$.

This result subsumes several result governing stationary sequences and curves, helices, stationary orthogonally-scattered and stationary quasi-isometric measures, due to Kolmogorov [13], Karhunnen [12], Schoenberg and von Neumann [21] and the writer[16, 17].

4. Propagators on involutory semi-groups

The theory of propagators becomes deeper and more interesting when Γ is an involutory s.g. (briefly *s.g.), i.e. Γ safisfies (2.1) (iv), and is also equipped with a function * on Γ to Γ such that

$$(4.1) \qquad \forall s,t \in \Gamma, \quad (s+t)^* = t^* + s^*, \quad t^{**} = t \ \& \ 0^* = 0.$$

This fruitful concept originated with Nagy in the 1950's [21:p.20]. It has wide scope: every group Γ is a *s.g. with $t^* \underset{d}{=} -t$, and every abelian s.g. Γ is a *s.g. with $t^* \underset{d}{=} t$. As Nagy observed [21: p.21] the notion of PD function on a *s.g. makes sense:

4.2. <u>Definition</u> a function $R(.)$ on a *s.g. Γ to $CL(W,W^*)$, where W is a Banach space, is called <u>positive definite</u> (PD), iff. the kernel $K(\cdot\ ,\ \cdot)$ defined by

$$\forall\ s,t \in \Gamma\ ,\quad K(s,t) \underset{d}{=} R(t^*+s)$$

is PD on $\Gamma \times \Gamma$ to $CL(W,W^*)$ in the sense of 2.4.

In the special case $W = \mathbb{F}$, Γ is a group and $t^* = -t$, this definition reduces to the classical one for a PD function. It will be useful to write

$$(4.3)\qquad \forall^* \text{s.g. } \Gamma\ ,\quad \Gamma_{0+} \underset{d}{=} \{t^* + t:\ t \in \Gamma\}\ ,$$

and to call Γ_{0+} the set of <u>non-negative elements</u> of Γ.

The full potentialities of the concept of a *s.g. in propagator theory have become apparent only very recently. When the s.g. Γ acting on Λ is involutory, it becomes convenient to widen the concept of propagator so as to include functions $S(.)$ for which the linear operator $S(t)$ is not necessarily continuous and to ask the following preliminary question before turning to Q. 3.5 :

(4.4) <u>Question.</u> Let the *s.g.Γ act on Λ and let $\mathfrak{X}(.)$ be a Hilbertian variety on Λ . What conditions on the covariance kernel $K_X(\cdot\ ,\ \cdot)$ of $X(.)$ are necessary and sufficient to ensure that a family $(S(t)\ :\ t \in \Gamma$ of closed linear operators $S(t)$ from $\mathfrak{Z}_X$ to $\mathfrak{S}_X$ such that

(i) $\forall\, t \in \Gamma,\ \mathfrak{D} \underset{d}{=} \bigcup_{\lambda\in\Lambda} X(\lambda)(W) \subseteq \mathfrak{D}_t \underset{d}{=} \text{dom.}\ \ S(t) \subseteq \mathfrak{Z}_X$

(ii) $\forall\, t \in \Gamma\ \&\ \forall\, \lambda \in \Lambda,\quad S(t).X(\lambda) = X(t\circledast\lambda)$

(iii) $\forall\, t \in \Gamma,\quad \text{Rstr.}_{\mathfrak{D}}\ S(t^*) \subseteq S(t)^*$?

Q. 4.4 is answered in the following theorem [18: 4.7] :

(4.5. <u>Main Theorem.</u> Let the *s.g. Γ act on Λ and let $X(.)$ be a Hilbertian variety on Λ . Then the following conditions are equivalent :

(α) $X(.)$ has a propagator $S(.)$ of the type described in 4.4.

(β) the covariance kernel $K_X(\cdot\ ,\ \cdot)$ of $X(.)$ has the <u>transfer property</u> :

$$\forall\, \lambda, \lambda' \in \Lambda \;\&\; \forall t \in \Gamma, \quad K_X(\lambda,\, t^*\oplus\lambda') = K_X(t\oplus\lambda, \lambda').$$

This theorem is an important stepping stone on the way to the answer of the more interesting question 3.5 for a *s.g. Γ. It is necessary, however, to study one of its corollaries, before turning to the answer, cf. [18: 4.8(c),(d)]:

4.6. <u>Corollary.</u> Let Γ ,Λ, $X(\cdot)$ be as in Thm. 4.5 and $K_X(\cdot\,,\cdot)$ have the transfer property 4.5(β). Then

(a) $\forall\, x \in \langle\mathfrak{D}\rangle \;\&\; \forall\, t \in \Gamma, \quad |S(t)x|^2 \leq |S(t^*+t)x|\cdot|x|$

(b) $\forall\, x \in \langle\mathfrak{D}\rangle\,,\; \forall\, p \in \Gamma_{0+} \;\&\; \forall\, n \in \mathbb{N}_+\,,$

$$|S(p)x|^{2^n} \leq |S(2^n p)x|\cdot|x|^{2^n-1}$$

Dividing in 4.6(b) by $|x|^{2^n}$, taking the 2^n th root and then letting $n \to \infty$, we obviously get

$$\forall\, x \in \langle\mathfrak{D}\rangle \setminus \{0\} \;\&\; \forall p \in \Gamma_{0+}\,, \quad \frac{|S(p)x|}{|x|} \leq \lim_{n\to\infty} |S(2^n p)x|^{2^{-n}}$$

Putting $p = t^* + t$ and using 4.6(a), we get

$$(4.7) \quad \forall\, x \in \langle\mathfrak{D}\rangle\setminus\{0\} \;\&\; \forall\, t\in\Gamma\,, \quad \frac{|S(t)x|^2}{|x|^2} \leq \lim_{n\to\infty} |S(2^n(t^*+t)x)|^{2^{-n}}$$

With the aid of these and similar results we are able to answer the basic question 3.5 for *s.g., cf. [18: 4.10] :

4.8. <u>Main Theorem.</u> II. Let the *s.g. Γ act on Λ . Then the following conditions are equivalent:

(α) $X(\cdot)$ has a propagator $S(\cdot)$ in the strict sense of Def. 3.1(b).

(β) The covariance kernel of $K_X(\cdot\,,\cdot)$ of $X(\cdot)$ has the transfer property 4.4(β) and is subject to the following inequality: $\exists$ a function $\gamma(\cdot)$ on Γ to $\mathbb{R}_{0+}$ such that

$$\forall \lambda \in \Lambda \;\&\; \forall\, t \in \Gamma, \quad 0 \leq K_X(t\oplus\lambda),\, t\oplus\lambda) \leq \gamma(t)\cdot K_X(\lambda,\lambda).$$

We shall refer to the last as the <u>mild translational inequality.</u> It is considerably simpler than the <u>Nagy translational</u> in 3.6, which in reality comprises an infinite sequence of inequalities, the rth one having r^2 terms on each side. In the mild inequality only the two 0th terms of these r^2 terms, so-to-speak, are compared.

5. Varieties on involutory semi-groups

Apart from the paper [18], the literature on the subject has dealt almost exclusively with the case where the s.g. Γ acts on Γ itself, i.e. where the variety X(.) has the domain Γ, or to put it differently, cf. 3.4, where the phase space Λ of the system is the semi-group Γ itself. This case is of course covered by our concept of Γ acting on Λ, for now $\Lambda = \Gamma$ and the $\oplus$ operation on $\Gamma \times \Lambda$ to Λ is just the + operation of the s.g. Γ.

When the s.g. Γ is not involutory, the case $\Gamma \mathrel{\overset{=}{\scriptscriptstyle d}} \Lambda$ does not lead to anything really new: we just get the results of §3 with + instead of $\oplus$. Of much greater interest is the case

$$(5.1) \qquad \Lambda = \Gamma = \text{a *s.d.}$$

This is because the deeper results of §4 are now at our disposal, and as noted at the outset of §4, the concept of a *s.g. has wide scope. For instance, a stationary random field or process is by definition a Hilbertian vector-valued variety parametrized on l.c.a. group Γ, the Hilbert space now being $L_2(\Omega, \mathcal{B}, P; \mathbb{F})$. Since every group Γ is a *s.g. with $t^* \mathrel{\underset{d}{=}} -t$, it follows that the theory of such fields and processes comes under the scope of (5.1). We should add, however, that there are important parts of the theory which go beyond the scope of (5.1), since $\Lambda \neq \Gamma$. Such is the case, for instance, with the propagator theory of helical varieties and of quasi-isometric measures, cf. Examples 3.3. For such cases we have to fall back on §§3, 4.

In this section we shall deal with the novel features which emerge when (5.1) prevails. In the first place, the important notion of the covariance <u>function</u> (as opposed to kernel) of a variety arises:

5.2. <u>Definition.</u> Let X(.) be a variety on a *s.g. Λ to $CL(W, \mathcal{H})$. We say that X(.) has a <u>covariance function</u> $R_X(\cdot)$, iff. $R_X(\cdot)$ is a function on Λ to $CL(W, W^*)$ such that

$$\forall\, \lambda, \lambda' \in \Lambda \quad R_X(\lambda'^* + \lambda) = K_X(\lambda, \lambda').$$

where $K_X(\cdot\,,\cdot)$ is the covariance kernel of X(.).

It follows of course from (2.5) & Def. 4.2 that the covariance function $R_X(\cdot)$, when it exists, is PD on Λ. Whereas all varieties X(.) on Λ possess a covariance kernel $K_X(\cdot\,,\cdot)$ on $\Lambda \times \Lambda$, not all varieties X(.) on a *s.g. Λ possess a covariance function

$R_X(\cdot)$ on Λ. In this regard we have the following theorem, which should be treated as the version for the case (5.1) of Thm. 4.5.

5.3. <u>Main Theorem</u> III. Let $X(\cdot)$ be a Hilbertian variety on the *s.a. Λ, and let Λ act on Λ. Then the following conditions are equivalent :

(α) $X(\cdot)$ possesses a covariance function $R_X(\cdot)$,

(β) the covariance kernel $K_X(\cdot\,,\,\cdot)$ of $X(\cdot)$ has the transfer property 4.5 (β),

(γ) $X(\cdot)$ has a closed, densely defined propagator $S(\cdot)$ of the type described in 4.4.

The equivalence (α) ⟺ (β) is easy to werify, and the equivalence (β) ⟺ (γ) is contained in Thm. 4.5.

It is clear from Thm. 5.3 that in the case (5.1) all we need for an affirmative answer to the basic question 3.5 is that $R_X(\cdot)$ should exist and the covariance kernel $K_X(\cdot\,,\,\cdot)$ should satisfy the mild translational inequality 4.8(β). It is easy to see that this simply means that $R_X(\cdot)$ should exist and satisfy the condition 5.4(β) below. But in the case (5.1) there also emerges a new condition, viz. 5.4(γ) below, which does not seem to have any analogue in the general setting of § 3,4. This was obtained recently by F.H. Szafraniec [27 : Prop.] on the basis of his own work and that of the Polish school. It is incorporated in our next result which should be treated as a consolidated version for the case (5.1) of Thm. 4.8, cf. [18 : 4.13] :

5.4 <u>Main Theorem</u> IV. Let $X(\cdot)$ be a Hilbertian variety on the *s.g. Λ and let Λ act on Λ. Then the following conditions are equivalent :

(α) $X(\cdot)$ has a propagator $S(\cdot)$ in the strict sense of Def. 3.1(b),

(β) $X(\cdot)$ has a covariance function $R_X(\cdot)$ subject to the following inequality: $\exists$ a function $\gamma(\cdot)$ on Λ to $\mathbb{R}_{0+}$ such that

$$\forall\ \lambda,t \in \Lambda,\quad 0 \leq R(\lambda^*+t^*+t+\lambda) \leq \gamma(t).R(\lambda^*+\lambda)$$

(γ) $X(\cdot)$ has a covariance function $R_X(\cdot)$ subject to the following inequality : $\exists$ a function $\sigma(\cdot)$ on Λ to $\mathbb{R}_{0+}$ and $\exists$ $c \in \mathbb{R}_+$ such that

$$\forall\ s,t \in \Lambda,\quad |R_X(t)| \leq c\cdot\sigma(t)\ \&\ \sigma(s+t) \leq \sigma(s).\sigma(t).$$

Proof. The equivalence $(\alpha) \Longleftrightarrow (\beta)$ is established in [18 : 4.13]. To complete the proof, we shall show that

$$(\alpha) \Longrightarrow (\gamma) \Longrightarrow (\beta).$$

Let (α) hold. Then $\forall\ t \in \Lambda$, $X(t) = S(t)X(0)$ and so

$$|R_X(t)| \underset{d}{=} |K_X(t,0)| = |X(0)^*S(t)X(0)| \leq |X(0)|^2.|S(t)|$$

With $c = |X(0)|^2$ and $\sigma(t) \underset{d}{=} |S(t)|$, we have (γ), since by (3.2),

$$|S(s+t)| \leq |S(s)|\ .|S(t)|.$$

Next let (γ) hold. Then $X(.)$ has a covariance function $R_X(.)$ and hence by Thm. 5.3, $K_X(\cdot\ ,\ \cdot)$ has the transfer property 4.5 (β). It follows from (4.7) that

(1) $\quad \forall\ x \in \mathfrak{Z}\ \&\ \forall t \in \Lambda,\quad |S(t)x\ |^2 \leq \varliminf_{n\to\infty} |S(2^n p)x|^{2^{-n}} \cdot |x|^2,$

where $p \underset{d}{=} t^* + t$. Now let $x \underset{d}{=} X(a)(w)$, $a \in \Lambda\ \&\ w \in W$. Then

(2) $\quad |x|^2 = |X(a)(w)|^2 = (X(a)w,\ X(a)\ w) = [R_X(a^*+a)w](w).$

Similarly,

(3) $\quad |S(t)x|^2\ \ |X(t+a)w\ |^2 = [R(a^*+t^*+t+a)w](w),$

and

$$\begin{aligned}|S(2^n p)x|^2 &= |X(2^n p+a)w|^2 = [R(a^*+2^{n+1}p+a)w](w)\\ &\leq |R(a^*+2^{n+1}p+a)|\,|w|^2, \quad \text{cf.}[18\text{: } 2.4(f)]\\ &\leq c\ .\sigma(a^*)\{\sigma(p)\}^{2^{n+1}}\ \sigma(a).|w|^2, \quad \text{by } (\gamma).\end{aligned}$$

Taking the 2^{n+1} root on both sides, and letting $n \longrightarrow \infty$, we get

(4) $\quad \varliminf_{n\to\infty} |S(2^n p)x|^{2^{-n}} \leq \sigma(p) \underset{d}{=} \sigma(t^*+t).$

Substituting in (1) from (2)-(4), we get $\forall\ a,t \in \Lambda$ and $\forall w \in W$,

$$[R(a^*+t^*+t+a)w](w) \leq \sigma(t^*+t).[R(a^*+a)w](w).$$

Hence, cf. (1.2),

$$\forall\ a,\ t \in \Lambda,\quad R(a^*+t^*+t+a) \leq \sigma(t^*+t).R(a^*+a).$$

Thus (β) holds with $\gamma(t) \underset{d}{=} \sigma(t^*+t)$.

Now a Banach algebra $\mathbb{A}$ over $\mathbb{F}$ with unit $\underline{1}$ and isometric involution * is of course a *s.g. under multiplication with neutral element $\underline{1}$, as well as a Banach space over $\mathbb{F}$ under addition. It turns out that if a function $R(\cdot)$ on $\mathbb{A}$ to $CL(W,W^*)$ is PD on (the *s.g.) $\mathbb{A}$ and linear on (the vector space) $\mathbb{A}$, then the mild translational inequality given in 5.4(β) automatically holds with $\gamma(t) = |t|^2$, $t \in \mathbb{A}$. It also turns out that any variety $X(\cdot)$ on $\mathbb{A}$ to $CL(W,\mathcal{H})$ having such a covariance function $R(\cdot)$ is itself a continuous linear operator on the Banach space $\mathbb{A}$. In this way we get from Thm. 5.4 the following theorem [18: 4.14] :

<u>5.5 Theorem.</u> Let (i) $\mathbb{A}$ be a Banach algebra over $\mathbb{F}$ with unit $\underline{1} \ni |1| = 1$ and an isometric involution *; (ii) $R(\cdot)$ be a linear operator on (the vector space) $\mathbb{A}$ to $CL(W,W^*)$, where W is a Banach space over $\mathbb{F}$; (iii) $R(\cdot)$ be a PD function on (the multiplicative *s.g.) $\mathbb{A}$ to $CL(W,W^*)$; (iv) $X(\cdot)$ be the variety on $\mathbb{A}$ to $CL(W,\mathcal{H})$ with covariance function $R(\cdot)$. Then

(a) $R(\cdot)$ is continuous on $\mathbb{A}$, $|R| = |R(\underline{1})|$, and

$$\forall a,t \in \mathbb{A} \ , \ 0 \leq R(a^*t^*ta) \leq |t|^2 R(a^*a) \ ;$$

(b) $X(\cdot) \in CL(\mathbb{A}, CL(W,\mathcal{H}))$ & $|X| = \sqrt{|R(\underline{1})|}$;

(c) $X(\cdot)$ possesses a propagator $S(\cdot)$ on $\mathbb{A}$ to $CL(\mathfrak{S}_X,\mathfrak{S}_X)$,

$$\text{i.e. } \forall a,t \in \mathbb{A}, \ X(ta) = S(t).X(a);$$

moreover, $\forall t \in \mathbb{A}$, $S(t^*) = S(t)^*$, and $S(\cdot)$ is a linear contraction of norm 1 on $\mathbb{A}$ to $CL(\mathfrak{S}_X,\mathfrak{S}_X)$; thus the propagator $S(\cdot)$ is a *-representation of $\mathbb{A}$ in $CL(\mathfrak{S}_X,\mathfrak{S}_X)$. ≠≠

This result is a variant of a theorem due to Stinespring [26 : Thm. 1], cf. [18: Thm. 4.15 & Remarks 4.16].

<u>6. The Kernel Theorem and dilations</u>

The ideas od <u>dilation</u> and <u>dilation theorem</u> have undergone expansion during the last decade or two. As originally conceived by Nagy and Halmos, cf. e.g. [10: p. 118], a simple dilation theorem asserts the embeddability of a Hilbert space W into a larger Hilbert space $\mathcal{H}$ so that a given operator R on W to W can be retrieved by projection from a simpler operator $\tilde{R}$ on $\mathcal{H}$ to $\mathcal{H}$; more precisely,

---

≠≠ i.e. $S(\cdot)$ is a continuous involution-preserving multiplicative linear operator on $\mathbb{A}$ to $CL(\mathfrak{S}_X, \mathfrak{S}_X)$.

(6.1) $RP = P\tilde{R}P$, equivalently $R = \mathrm{Rstr.}_W P\tilde{R}$,

where P is the orthogonal projection on $\mathcal{H}$ onto W, and "$\mathrm{Rstr.}_W$" stands for "restriction to W". $\tilde{R}$ is then spoken of as <u>dilation of R</u> Less simple dilation theorems assert the same type of embeddability for all positive powers R^n, $\tilde{R}^n$.

This concept of dilation emerges from the traditional identification of the given Hilbert space W with a subspace $\tilde{W}$ of another (related) Hilbert space $\mathcal{H}$ to which W is isometrically isomorphic, i.e. from the replacement of the accurate relation $W \simeq \tilde{W} \subseteq \mathcal{H}$ by the more artificial and supposedly simpler relation $W = \tilde{W} \subseteq \mathcal{H}$. The recent trend has been to avoid the identification of $\tilde{W}$ and W, and to adopt the definition of dilation which then emerges, viz:

<u>6.2. Definition</u>. (a) Let W, $\mathcal{H}$ be Hilbert spaces over the field $\mathbb{F}$, R be any operator (not necessarily linear) from W to W and $\tilde{R}$ be any operator from $\mathcal{H}$ to $\mathcal{H}$. We say that <u>$\tilde{R}$ is a dilation of R</u>, iff. $\exists$ a linear isometry J on W into $\mathcal{H}$ such that $R = J^*\tilde{R}J$.

(b) Let Λ be any set, W and $\mathcal{H}$ be Hilbert spaces over $\mathbb{F}$, $R(\cdot)$ be a W-to-W operator-valued function on Λ, and $\tilde{R}(\cdot)$ be a $\mathcal{H}$-to-$\mathcal{H}$ operator-valued function on Λ. We say that <u>$\tilde{R}(\cdot)$ is a dilation of $R(\cdot)$</u>, iff. $\exists$ a linear isometry J on W into $\mathcal{H}$ such that

$$\forall \lambda \in \Lambda, \quad R(\lambda) = J^*\tilde{R}(\lambda)J.$$

In this approach the equation (6.1) gives way to that in 6.2(a). A dilation theorem is one that claims for a given type of W-to-W operator R or W-to-W operator-valued function $R(\cdot)$, the existence of a dilation $\tilde{R}$ of $R(\cdot)$ of a specyfic type, in the sense of Def. 6.2. It is possible to adopt a broader definition of dilation by allowing the J in 6.2 to be any continuous linear (not necessarily isometric) operator on W to $\mathcal{H}$. This in essence is what is done in Stinespring's work [26: Thm. 1]. This reinterpretation of J in turn allows us to consider the W in 6.2 to be any Banach (not just Hilbert) space. This has been done by Górniak & Weron [31],[30]. Thus there looms the possibility of a dilation theory of considerably wider scope then the one in existence. In this paper, however, we shall as in [18] abide by Def. 6.2.

When dilation theory is based on Def. 6.2 it is seen to be just another facet of propagator theory.

This connection has been obscured by adherence to the narrow conception of dilation given in (6.1), by the neglect of the Kernel Theorem in the literature on dilations, and by the usual enunciation of the latter with emphasis on the reproducing-kernel property rather than on the covariance property. It is therefore important to formulate the Kernel Theorem appropriately as follows cf. [18 : 2.10]:

6.3. Kernel Theorem. (Kolmogorov, Aronszajn, Pedrick)*. Let Λ be any set and W be a Banach space over $\mathbb{F}$ with adjoint W*. Then every PD Kernel K(. , .) on $\Lambda \times \Lambda$ to CL(W,W*) is the covariance kernel of a Hilbertian variety X(.) on Λ.

More fully, for all PD kernels K(. , .) on $\Lambda \times \Lambda$ to CL(W,W*), $\exists$ a cardinal no. α such that for all Hilbert spaces $\mathcal{H}$ with dim $\mathcal{H} \geqslant \alpha$, $\exists$ a variety X(.) on Λ to CL(W,$\mathcal{H}$) for which $K(\cdot\,,\cdot) = X_K(\cdot\,,\cdot)$, i.e.

$$\forall\ \lambda,\lambda' \in \Lambda, \quad X(\lambda')^*X(\lambda) = K(\lambda,\lambda').$$

The Kernel Thm. 6.3 enables us to easily assert the following simple result which reveals the nexus between dilations and propagators [18: 5.1] :

6.4. General Dilation Theorem. Let

(i) Λ be a semi-group under + (not necessarily abelian) with neutral element 0,

(ii) W be a Hilbert space over $\mathbb{F}$,

(iii) K(. , .) be a PD Kernel on $\Lambda \times \Lambda$ to CL(W,W) for which $K(0,0) = I_W$.

Then

(a) if a variety X(.) on Λ to CL(W,$\mathcal{H}$) the covariance kernel of which is K(. , .),$\mathcal{H}$ being as in 6.3, has the propagator S(.), then S(–)*.S(.) is the dilation of K(. , –) in the Hilbert space $\mathfrak{S}_X \subseteq \mathcal{H}$;

(b) if S(.) is a semi-group on Λ to CL($\mathcal{H}$,$\mathcal{H}$) and S(–)*.S(.) is a dilation of K(. , –), then a restriction $S_0(\cdot) \subseteq S(\cdot)$ is the propagator of a variety X(.) having K(. , .) as its covariance kernel.

---

* This theorem subsumes a wide variety of results, some of which were found later by others, e.g. Vakhania [28] and Chobanian [6], who were unaware of Pedrick's unpublished paper [23].

This theorem enables us to easily recast theorems on propagators (3-5) into dilation theorems. We get in this way the following simplified and strengthened versions of Nagy's Principal Thm. [21 p. 21] and of Stinespring's Thm. [26: Thm. 1] from our Thms. 5.4 and 5.5, cf. [18: 5.3, 5.4].

6.5 <u>Theorem</u>. (Simplified Nagy Thm.) Let (i) Λ be an involutory (additive but not necessary abelian) semi-group with neutral element 0; (ii) W be a Hilbert space over $\mathbb{F}$; (iii) R(.) be a PD function on Λ to CL(W,W), cf. 4.1(b), such that $R(0) = I_W$, which satisfies either the mild translational inequality 5.4(β) or the Szafraniec inequality 5.4(γ). Then

(a) $\exists$ a Hilbert space $\mathcal{H}_0$ over $\mathbb{F}$, $\exists$ a linear isometry J on W to $\mathcal{H}_0$ & $\exists$ a semi-group S(.) on Λ to CL($\mathcal{H}_0$, $\mathcal{H}_0$)

$$\forall \lambda \in \Lambda, \quad R(\lambda) = J^*S(\lambda)J,$$

i.e. R(.) has the dilation S(.) in $\mathcal{H}_0$. Here $\mathcal{H}_0 = \mathfrak{S}_X$, J = X(0) and S(.) is the propagator of X(.), where X(.) is a variety on Λ having the covariance function R(.), cf. 6.3 & 5.2;

(b) $\quad \forall \lambda \in \Lambda, \quad S(\lambda^*) = S(\lambda)^* \;\&\; |S(\lambda)| < \sqrt{\gamma(\lambda)}.$

6.6. <u>Theorem</u> (Stinespring). Let (i) IA be a Banach algebra over IF with unit $\underline{1} \ni |\underline{1}| = 1$ and an isometric involution *, (ii) R(.) be a linear operator on (the vector space) IA to CL(W,W) where W is a Hilbert space over IF, (iii) R(.) be a PD function on (the *s.g.) IA to CL(W,W) & $R(\underline{1}) = I_W$. Then

(a) $\exists$ a Hilbert space $\mathcal{H}_0$, $\exists$ a *-representation S(.) of IA in CL($\mathcal{H}_0$, $\mathcal{H}_0$), and $\exists$ a linear isometry J on W to $\mathcal{H}_0$ such that

$$\forall a \in \text{IA}, \quad R(a) = J^*S(a)J.$$

This S(.) is a dilation of R(.) in $\mathcal{H}_0$.

(b) In particular,(a) holds when IA is an abelian C*-algebra with unit $\underline{1}$ and R(.) is any positive linear operator on IA to CL(W,W).

The utility of these powerful theorems in the actual determination of the dilation of a specific operator-valued function R(.) rests on the possibility of associating with R(.) a non-trivial kernel K(.,.) the positive definiteness of which hinges on the specific features of R(.) . Such an association is not always obvious. The important work of Nagy on contractions, and of Foias,

Lebow, Arveson, Mlak and others on spectral sets, Silov boundaries, etc. may be regarded as methods of affecting such associations, cf. [18 : 5.5 & remarks]. For instance, for the s.g. (C^n : n IN$_{0+}$) of linear contractions on a Hilbert space W, Nagy defined the kernel K(.,.) on IN x IN by :

$$K(m,n) = C(m-n), \quad C(n) \underset{d}{=} \begin{cases} C^n & n \geq 0 \\ (C^{-n})^* , & n \geq 0. \end{cases}$$

The non-obvious fact that this K(. , .) is PD on the C^n being contractions, cf. [21: p. 30], [22 : p.29]. But once such an association has been affected, we can appeal to the theory of propagators. We may therefore lay down the following routine for obtaining dilations :

6.7 <u>Procedure.</u> Given a Hilbert space W and a W-to-W operator-valued function R(.) on Λ, (i) associate with it an appropriate PD kernel K(. , .); (ii) check if this K(. , .) satisfies the conditions for the existance of a propagator for the Hilbertian variety X(.) having covariance K(. , .); (iii) if so, determine the propagator S(.). Then S(.) is the dilation of R(.).

In this the hardest step is (i), for once the appropriate PD kernel is known we can fall back on results on propagators, of the type given above, for the reamaining steps. In some problems it is illuminating to supplement the step (ii) by determining the Hilbertian variety X(.) having the covariance K(. , .), even though it plays no role in the final analysis. This is the case, for instance, in the Naimark Dilation Thm. for W-to-W, non-negative hermitian operator-valued measures. Here the associated variety is a <u>quasi-isometric- measure</u>, [17]. It follows from the theory of such measures, without any appeal to the results of § 3-5, that its so-called <u>spatial spectral measure</u> is the dilation, cf. [18: 5.6-5.13].

Nagy observed that his Principal Theorem could be used to prove the Halmos Theorem on normal extensions. Now a transition analogous to the one from (6.1) to 6.2 for dilations can be affected for the related concept of <u>extension</u>, cf. [18: 1.5]. As shown in [18: §6] Our strengthened version 6.5 of the Nagy Theorem can then be employed to prove the strenthened version due to Bram [5 : Thm. 1] of the Halmos Theorem on normal extensions.

7. Bibliographical remarks and new developments.

We owe to Halmos and Nagy (c. 1950) the concept of dilation, according to which the dilation of W-to-W operator-valued function R(.) on Λ, (W = a Hilbert space) is a $\mathcal{H}$-to-$\mathcal{H}$ operator valued function R(,) on Λ, where $\mathcal{H}$ is a superspace of W and R(.).P = P.$\tilde{R}$(.).P,P beining the orthogonal projection on $\mathcal{H}$ onto W. Thereafter, Stinespring [26], Lebov [4], Mlak [19] and others have widened the scope of the concept by relaxing the requirements that $\mathcal{H}$ be a superspace of W, and P the projection on $\mathcal{H}$ onto W. This trend, continued in the present paper, is pushed perhaps farthest in the 1976 paper [31] of Górniak & Weron, where in W is a Banach space and the only stipulation is that there exist a W-to-$\mathcal{H}$ continuous linear operator J such that R(.) = J*.$\tilde{R}$(.)J.

The existence of a nexus between the widened concept of dilation (6.2) and the propagators S(.) of Hilbertian varieties X(.) is implicit in Getoor's paper [7], 1956. But the fusion of these ideas into a unified theory in which the semi-group Γ is distinguished from the space Λ on which it acts, and the Kernel Theorem is systematically utilized seem to be due to the writer [18]*.

The possibility that the Nagy inequality [21: p.21, (c)] can be mitigated when $\Lambda = \Gamma$ = an involutory s.g. with neutral element has loomed on the horizon ever since Bram's important improvement of the Halmos theorem on normal extensions [5],(1955). The first successful move in this direction seems to be that of W. Arveson for a bounded PD function (c. 1971, unpublished); it required an appeal to Stinespring's Theorem [26: Thm. 1] on C* algebras. The idea that mitigation of the Nagy inequality is possible also for unbounded PD functions and kernels, and is justifiable without recourse to C* algebras, occurred to Szafraniec and to the writer independently in 1976, cf [27], [18]. It should be noted, however, that although the conditions due to Szafraniec and the writer are equivalent when $\Lambda = \Gamma$ = a *s.g. with neutral element, cf. 5.4(β)(γ), the former's condition on PD functions R(.) has no analogue for kernels K(. , .) when $\Lambda \neq \Gamma$, whereas the latter's condition has, cf. 4.8(β). It therefore seems that our Thm. 4.8 it to-date the

*This paper is a revised and substantially enlarged version of a lecture delivered at the Symposium on "Operator Theory of Networks and Systems" in Montreal, Canada, in August, 1975, entitled "An explicit treatment of dilation theory".

most potent result covering all situations involving *s.g.'s with a neutral element.

It is of course necessary to extend these considerations to *s.g.'s devoid of a neutral element. Here we must refer to the efforts in this direction of several Polish mathematicians, cf. [20 & references therein].

Two other directions in which these developments can be extended appear in an unpublished paper by Shonkwiler and Faulkner [25]. They adopt the superspace concept of dilation (6.1) and fuse the two issues in their treatment, For clarity we shall comment on each individually, using the framework developed above.

(1) For any linear manifold W_o of a vector space W adopt the notation

$$(W_o) \underset{d}{=} \{ T: T \text{ is a linear operator from } W \text{ to } W \text{ with domain } \mathfrak{D}_T \supseteq W_o \}.$$

Now let W be a Hilbert space, W_o be a linear manifold everywhere dense in W, and Λ be an arbitrary set. The authors define the concept of a PD kernel $\underline{K(\cdot\,,\,\cdot) \text{ on } \Lambda \times \Lambda \text{ to } \mathcal{L}(W_o)}$. The values $K(\lambda,\lambda')$ are not necessarily continuous operators, and this brings up the following interesting question :

$$(7.1)\begin{cases}\text{In what form does the Kernel Thm. 6.3 survive}\\ \text{for such (not necessarily continuous) linear-operator-valued}\\ \text{PD kernels on } \Lambda \times \Lambda \text{ ?}\end{cases}$$

Clearly the Hirbertian varieties $X(\cdot)$ having such covariance kernels must themselves have values which are discontinuous linear operators from W to $\mathcal{H}$. Such varieties deserve further study because of their natural appearance in other situations. For instance, they are required in order to secure the completeness of certain L_2 spaces, which are related to the Naimark Dilation Theorem but also have a much broader significance, cf. [18: 5.13] & [17: 9.9] & [15].

(2) In [25] the authors obtain a generalization of the Nagy Principal Theorem by relaxing the requirement that $\Lambda = \Gamma =$ a *s.g. having a neutral element 0. They merely require that $\Lambda = \Gamma$ is any set endowed with an involution * and a "distinguished element" 0 such that $0^* = 0$. A simple example is the set $\Gamma = (\mathbb{C} \setminus \mathbb{R}) \cup \{0\}$, the * being complex conjugation, which appears naturally in resolvent theory, cf. [24]. They prove a general dilation theorem for kernels $K(\cdot\,,\,\cdot)$ on $\Lambda \times \Lambda$ of the form

$$(7.2)\qquad K(\lambda,\lambda') = \int_{\Omega} F(\lambda,\lambda',\omega)\,\mu(d\omega),$$

where $(\Omega,\mathfrak{B},\mu)$ is a measure space and $F(\cdot,\cdot,\cdot)$ is a function on $\Lambda^2 x\,\Omega$ to $CL(W,W)$, W = a Hilbert space. Nagy's own theorem emerges as a corollary when $\Lambda = \Gamma$ is a *s.g. with neutral element upon suitable choice of $(\Omega,\mathfrak{B},\mu)$ and $F(\cdot,\cdot,\cdot)$. The scope and kinematical significance of such involution-bearing sets are as yet unclear but provoke interest.

As indicated above, the authors merge the issues (1) and (2) in their main theorem [25: §1] by requiring that the $F(\cdot,\cdot,\cdot)$ in (7.2) take its values in $\mathcal{L}(W_0)$. They bypass the basic question (7.1) in which Λ need possess neither involution nor distinguished element.

References

[1] N. Aronszajn, La theorie generale de noyaux reproduisants et ses applications, Proc.Cambridge Philos. Soc. 39 (1944), 133-153.

[2] N. Aronszajn, Theory of reproducing kernels, Trans.Amer. Math. Soc. 68 (1950), 337-404.

[3] W. Arveson, Subalgebras of C*-algebras, Acta Math. 123 (1969), 141-224.

[4] W. Arveson, Subalgebras of C*-algebras II, Acta Math. 128 (1972), 271-308.

[5] J. Bram, Subnormal operators, Duke Math. J. 22 (1955), 75-94.

[6] S.A. Chobanian, On a class of functions of a Banach space valued stationary stochastic process, Sakharth SSR Mecn. Acad.Moambe 55 (1969), 21-24.

[7] R.K. Getoor, The shift operator for non-stationary stochastic processes, Duke Math. J. 23 (1956), 175-187.

[8] J. Hadamard : Lectures on Cauchy's Problem in linear differential equations, Dover Publications, New York, 1952.

[9] P.R. Halmos, Normal dilations and extensions of operators, Summa Brasiliensis Math. 2 (1950), 125-134.

[10] P.R. Halmos, A Hilbert space problem book, van Nostrand, New York, 1967.

[11] E. Hille and R.S. Phillips, Functional analysis and semi-groups, Amer. Math. Soc. Colloq. Publ., Vol. 31, Amer. Math. Soc., Providence, R.I., 1957.

[12] K. Karhunen, Uber lineare Methoden in der Wahrscheinlich-keitsrechnung, Ann. Acad. Sci. Fenn. Ser. A.1 37 (1947)

[13] A.N. Kolmogorov, Stationary sequences in Hilbert space, Bull. Math. Univ. Moscow 2 (1941), 1-40.

[14] A. Lebow, On von Neumann's theory of spectral sets, J. Math. Anal. & Appl. 7 (1963), 64-90.

[15] V. Mandrekar & H. Salehi, The square-integrability of operator-valued functions with respect to a non-negative operator-valued measure and the Kolmogorov isomorphism theorem, Indiana Univ. Math. J. 20 (1970), 545-563.

[16] P. Masani,Orthogonally scattered measures, Adv. in Math. 2 (1968), 61-117.

[17] P. Masani, Quasi-isometric measures and their applications, Bull. Amer. Math. Soc. 76 (1970), 427-528.

[18] P. Masani, Dilations as propagators of Hilbertian varieties, S.I.A.M. J Anal. 9 (1978) to appear.

[19] W. Mlak, Operator inequalities and related dilations, Acta Sci. Math., 34 (1973),273-278.

[20] W. Mlak, Dilations of Hilbert space operators (General Theory) to appear in Dissertationes Math. 158

[21] B. Sz.-Nagy, Extensions of linear transformations in Hilbert space which extend beyond this space, F. Ungar, New York, 1960. (German Ed. Deutscher Verlag der Wissenchaften Berlin, 1950).

[22] B. Sz,-Nagy & C. Foias, Harmonic analysis of operators on Hilbert space, North Holland, New York, 1970.

[23] G.B. Pedrick, Theory of reproducing kernels in Hilbert spaces of vector valued functions, Univ. of Kansas Technical Report 19 (1957).

[24] R. Shonkwiler, On generalized resolvents and an integral representation of Nevanlinna, J. Math. Anal. & Appl. 40 (1972), 723-734.

[25] R. Shonkwiler & G. Faulkner, Kernel dilation in reproducing kernel Hilbert space and its applications to moment problems (to appear).

[26] W. Stinespring, Positive functions on C*-algebras, Proc.Amer. Math. Soc. 6(1955), 211-216.

[27] F.H. Szafraniec, Dilations on involution semi-groups Proc. Amer. Math. Soc. (to appear).

[28] N.N. Vakhania, The covariance of random elements in a Banach space, Thbilis Sanelmc. Univ. Gamoqeneb. Math. Inst.2 (1969) 179-184.

[29] J. von Neuman & I.J. Schoenberg, Fourier integrals and metric geometry, Trans. Amer. Math. Soc. 50 (1941), 226-251.

[30] A. Weron, Remarks on positive definite operator valued functions in Banach Spaces, Bull. Acad. Polonaise Sci. 24 (1976), 873-876.

[31] A. Weron and J. Górniak, An analogue of Sz.-Nagy's dilation theorem, Bull. Acad. Polonaise Sci. 24 (1976), 867-872.

University of Pittsburgh,
Pittsburgh, Pa, 15260
U S A

ON MINIMALITY OF INFINITE DIMENSIONAL STATIONARY STOCHASTIC PROCESSES

A.G. Miamee and H. Salehi

Abstract

The minimality problem for infinite dimensional stationary stochastic processes is studied, and spectral characterizations for minimal full rank processes are given. This work extends an earlier result of P. Masani on finite dimensional case.

A closed subset $\overline{H}$ or H is called a subspace if $\overline{H}$ is closed under the usual addition and $A\xi \in \overline{H}$ for any matrix A and any $\xi \in \overline{H}$ for which $A\xi$ is defined. We say $\xi \perp \eta$ if $(\xi, \eta) = 0$.

For any $B \subset H$ and $\overline{B} \subset \overline{H}$ we write

(a) $\mathfrak{S}(B)$ = subspace of H spanned by B,

(b) $\overline{\mathfrak{S}}(\overline{B})$ = subspace of $\overline{H}$ spanned by $\overline{B}$,

(c) $\mathfrak{S}(\overline{B})$ = subspace of H spanned by the coordinates of all vectors of $\overline{B}$, and

(d) $\overline{\mathfrak{S}}(B)$ = the set of all vectors in $\overline{H}$ with coordinates being in B.

It is easy to see that if $\overline{B}$ is a subspace of $\overline{H}$ then we have $\overline{B} = \overline{\mathfrak{S}}(\mathfrak{S}(\overline{B}))$. Hence $\overline{B} = \overline{\mathfrak{S}}(B)$ for some subspace B of H. Let ξ be an element of $\overline{H}$. we define, $(\xi|\overline{B})$, its projection on a subspace $\overline{B}$ of $\overline{H}$ to be the vector whose k-th coordinate is given by

$$(\xi|\overline{B})^k = (\xi^k|B).$$

2.1. Definition. A sequence ξ_n, $-\infty < n < \infty$ of elements of $\overline{H}$ is called infinite dimensional stationary stochastic process (S.S.P.) if the Gramian (ξ_m, ξ_n) depends only on $m - n$. It is easy to see that given any infinite dimensional S.S.P., say ξ_n, there exists an infinite dimensional nonnegative matrix valued measure defined on the unit circle such that

$$(\xi_m, \xi_n) = \frac{1}{2\pi}\int_0^{2\pi} e^{-i(m-n)\theta}\, dF(\theta)$$

dF is called the spectral measure of our process and in case that it is of the form $dF(\theta) = f(\theta)d\theta$, f is called the spectral density of the S.S.P.

2.2. <u>Definition</u>. A stationary stochastic process ξ_n, $-\infty < n < \infty$ is called minimal if for each integer j, $\xi_j \notin \overline{M}(j) = \overline{\mathfrak{S}}\{\xi_i,\ i \neq j\}$ or equivalently $\xi_j^k \notin M_{(j)} = \mathfrak{S}\{\xi_i,\ i \neq j\}$ for some k.

The two sided innovation process of ξ_n is defined by $\zeta_n = \xi_n - (\xi_n | \overline{M}(n))$, $-\infty < n < \infty$. We write $\Sigma = (\zeta_0, \zeta_0)$ and call it the two sided error matrix. Clearly a S.S.P. ξ_n is minimal if and only if $\Sigma \neq 0$. We call a S.S.P. to be minimal of full rank if $\Sigma \geqslant \lambda I$ for some positive number λ.

Following [5] for a nonnegative infinite dimensional matrix valued measure μ on the Borel subset of the unit circle, we define $L_2'(\mu)$ to be the collection of all row vector valued functions $g = \{g_k\}_{k=0}^{\infty}$ such that $g_k = 0$ for all except finitely many integers k, and for these k's, g_k is a trigonometric polynomial. In $L_2'(\mu)$ we introduce the following norm and inner product :

$$||g||_\mu = [\frac{1}{2\pi}\int_0^{2\pi} \sum_{k,l=1}^{\infty} g_k(e^{i\theta})\overline{g_l(e^{i\theta})d\mu_{kl}(\theta)}]^{1/2},$$

$$(g,h)_\mu = \frac{1}{2\pi}\int_0^{2\pi} \sum_{k,l=1}^{\infty} g_k(e^{i\theta})\overline{h_l(e^{i\theta})d\mu_{kl}}(\theta).$$

We identify two elements of $L_2'(\mu)$ whenever their difference has zero norm. We denote by $L_2(\mu)$ the Hilbert space obtained from completing $L_2'(\mu)$.

2.3 <u>Definition.</u> Let ξ_n be a S.S.P. with spectral density f. We define $L_2(f)$ to be the Hilbert space $L_2(\mu)$, where $\mu_{kl}(A) = \int_A f_{kl}(\theta)d\theta$. We denote the inner product of $L_2(f)$ by $(g,h)_f$ which can be written as

$$(g,h)_f = \frac{1}{2\pi}\int_0^{2\pi} g(e^{i\theta})f(e^{i\theta})h^*(e^{i\theta})d\theta .$$

Consistent with our earlier notations we write $\overline{L}_2(f)$ to be the set of all infinite dimensional matrix valued functions each row of which being in $L_2(f)$. We then give $\overline{L}_2(f)$ the row wise convergent topology, i.e. for Φ_n, $\Phi \in \overline{L}_2(f)$ we say $\Phi_n \longrightarrow \Phi$ in $\overline{L}_2(f)$ if and only if $\Phi_n^k \longrightarrow \Phi^k$ in $L_2(f)$ for each positive integer k, where Φ_n^k and Φ^k are the k-th row of Φ_n and Φ, respectively. For any Φ and Ψ in $\overline{L}_2(f)$ we write $(\Phi,\Psi)_f$ to mean the matrix whose (i,j)-th entry is $(\Phi^i,\Psi^j)_f$. It is easy to verify that for Φ, Ψ

in $\overline{L}_2(f)$, $(A\Phi, B\Psi)_f = A(\Phi,\Psi)_f B^*$ whenever $A\Phi$ and $B\Psi$ are in $\overline{L}_2(f)$. When $f = I$ we write L_2 and $\overline{L}_2$ instead of $L_2(f)$ and $\overline{L}_2(f)$, respectively.

2.4 Definition. Let ξ_n, $-\infty < n < \infty$ be a S.S.P. with spectral density f, then

(a) We define an isometric isomorphism S from $L_2(f)$ onto $H_\xi(\infty) = \mathfrak{S}\{\xi_n, -\infty < n < \infty\}$ by the mapping $\xi_n^k \longrightarrow \Psi_n^k$, where Ψ_n^k is the row vector $\{e^{-in\theta}\delta_{kl}\}$, through linearity.

(b) We define the map $\overline{S}$ as the infalion of S defined on $\overline{H}_\xi(\infty) = \overline{\mathfrak{S}}\{\xi_n, -\infty < n < \infty\}$ onto $\overline{L}_2(f)$ by the relation $(\overline{S}(\xi))^k = S(\xi^k)$ for each $\xi \in \overline{H}_\xi(\infty)$ and each positive integer k.

The following properties of $\overline{S}$ are trivial.

(a) $(\xi, \eta) = (\overline{S}\xi, \overline{S}\eta)_f$,

(b) S is a one-to-one continuous additive transformation.

(c) $\overline{S}(A\xi) = A\overline{S}(\xi)$, whenever $A\xi$ is defined.

In the rest of this paper we assume that our S.S.P. has a density f satisfying $0 < m(e^{i\theta})I \leq f(e^{i\theta}) \leq M(e^{i\theta})I$ for almost every θ. As noted in [5] in this case $L_2(f)$ consists of only l^2 valued functions. In fact $L_2(f)$ consists of all l^2 valued functions $g(e^{i\theta}) = \{g_k(e^{i\theta})\}_{k=1}^{\infty}$ with measurable entries such that $g(e^{i\theta})f(e^{i\theta})g^*(e^{i\theta}) = \lim_{N\to\infty} \sum_{k,l=1}^{N} g_k(e^{i\theta})f_{kl}(e^{i\theta})\overline{g_l(e^{i\theta})}$ exists for almost every θ and the resulting function is summable.

We now introduce the following useful definition.

2.5 Definition. Let g and h be row vector valued functions with measurable entries. We define $< g,h >_f$ to be

$$< g,h >_f = \frac{1}{2\pi}\int_0^{2\pi} g(e^{i\theta})f^{-1}(e^{i\theta})h^*(e^{i\theta})d\theta ,$$

whenever this integral exists and is finite. We denote by $H_2(f)$ the class of all row vector valued functions g for which $< g,g >_f$ is defined.

The followings are easy to verify :

(a) $g \in H_2(f)$ if and only if $gf^{-1/2} \in L_2$

(b) g and h in $H_2(f)$ implies that $g + h \in H_2(f)$.

The following lemma establishes an isometric isomorphism between $H_2(f)$ and $L_2(f)$

2.6 __Lemma.__ Let f be the spectral density of some S.S.P. Then the mapping $Tg = gf$ is an isometric isomorphism from $L_2(f)$ onto $H_2(f)$. Hence $L_2(f)$ being a Hilbert space implies that $H_2(f)$ is also a Hilbert space.

__Proof.__ For any $g,h \in L_2(f)$ we have

$$< Tg,Th >_f = \frac{1}{2\pi}\int_0^{2\pi}(g(e^{i\theta})f(e^{i\theta}))f^{-1}(e^{i\theta})(h(e^{i\theta})f(e^{i\theta}))^*d\theta$$

$$= \frac{1}{2\pi}\int_0^{2\pi} g(e^{i\theta})f(e^{i\theta})h^*(e^{i\theta})d\theta = (g,h)_f.$$

Hence T is an isometric isomorphism. For the ontoness we note that for any $g \in H_2(f)$ we have $T(gf^{-1}) = g$ and $gf^{-1} \in L_2(f)$.

2.7 __Definition.__ We denote by $\overline{H}_2(f)$ the space of all infinite dimensional matrix valued functions each row of which is in $H_2(f)$. For Φ and Ψ in $\overline{H}_2(f)$ we define $<\Phi,\Psi>_f$ to be the following infinite dimensional matrix

$$< \Phi,\Psi >_f = [< \Phi^i, \Psi^j >]_{i,j=1}^{\infty} .$$

Let the transformation $\overline{T}$ from $\overline{L}_2(f)$ into $\overline{H}_2(f)$ be the inflation of T. The results of Lemma 2.6 and the usual techniques can be used to show that $\overline{T}$ is a one-to-one transformation on $\overline{L}_2(f)$ onto $\overline{H}_2(f)$ which is an isometry. In fact for all Φ and Ψ in $\overline{L}_2(f)$ we have

$$< \overline{T}\Phi, \overline{T}\Psi >_f = (\Phi, \Psi)_f .$$

The following interesting result is due to P.Masani [3]. It gives a characterization for a minimal full rank finite dimensional S.S.P. We will give two natural extensions of this result for the infinite dimensional case.

2.8. __Theorem.__ Let ξ_n, $-\infty < n < \infty$ be a finite dimensional S.S.P. with spectral distribution F and the two sided predictor error matrix Σ.
Then ξ_n, $-\infty < n < \infty$ is minimal full rank if and only if F'^{-1}

exists a.e. and F'^{-1} is summable. In this case we have

$$\Sigma = [\frac{1}{2\pi}\int_0^{2\pi} F'^{-1}(e^{i\theta})d\theta]^{-1} .$$

For any stationary stochastic process ξ_n, $-\infty < n < \infty$. The L-dimensional subprocess $\xi_{L,m}$, $-\infty < n < \infty$ is defined via

$$(\xi_{L,n})^i = \begin{cases} \xi_n^i, \text{ if } 0 < i \leq L \\ 0 , \text{if } \quad i > L . \end{cases}$$

3. <u>Main results.</u> In this section we give some characterizations for minimality of infinite dimensional stationary stochastic processes which give rise to natural extensions of Theorem 2.8 due to Masani. In order to establish these results we need the following important lemma.

3.1 <u>Main lemma.</u> Let ξ_n, $-\infty < n < \infty$ be a S.S.P. such that (ξ_0, ξ_0) is a bounded operator on l_2. Let Σ and Σ_L be the two-sided predictor error matrices of ξ_n and $\xi_{L,n}$, respectively. Then $\Sigma \geq \lambda^2 I$ if and only if $\Sigma_L \geq \lambda^2 I_L$ for all positive integers L (I_L is the $L \times L$ identity matrix).

<u>Proof.</u> If $\Sigma \geq \lambda^2 I$ then clearly $\Sigma_L \geq \lambda^2 I_L$ for all $L > 0$. To prove the other way, let us assume $\Sigma_L \geq \lambda^2 I_L$ for all $L > 0$, and suppose $\Sigma < \lambda^2 I$, i.e. suppose there exists a sequence c_n, $-\infty < n < \infty$ with $\sum_{i=1}^{\infty} |c_i|^2 = 1$ such that

$$(1) \qquad \sum_{i,j=1}^{\infty} c_i \Sigma_{ij} \overline{c_j} = \lambda'^2 < \lambda^2$$

Let $\varepsilon = \lambda - \lambda'$ and take $N_1 > 0$ such that

$$(2) \qquad \sum_{j=n}^{\infty} |c_j|^2 < \frac{\lambda^2 - (\lambda' + \varepsilon/2)^2}{\lambda^2} , \text{ for all } n > N_1 .$$

We have

$$\|\sum_{i=1}^{\infty} c_i\xi_0^i\|^2 = (\sum_{i=1}^{\infty} c_i\xi_0^i, \sum_{i=1}^{\infty} c_i\xi_0^i) = \sum_{i,j=1}^{\infty} c_i(\xi_0^i,\xi_0^j)\overline{c}_j < \infty ,$$

because (ξ_0,ξ_0) is bounded. Hence there exists $N_2 > 0$ such that

(3) $$||\sum_{i=n}^{\infty} c_i\xi_0^i|| < \varepsilon/4, \text{ for all } n > N_2 .$$

If ζ_0 is the two sided linear predictor of ξ_n, $-\infty < n < \infty$ then (1) means

$$\lambda^2 > \lambda'^2 = \sum_{i,j=1}^{\infty} c_i \Sigma_{ij} \overline{c_j} = \Sigma_{ij} c_i(\zeta_0,\zeta_0)_{ij} \overline{c_j}$$

$$= ||\sum_{i=1}^{\infty} c_i \zeta_0^i||^2 = ||\sum_{i=1}^{\infty} c_i\xi_0^i - (\sum_{i=1}^{\infty} c_i\xi_0^i| M(0))||^2.$$

Hence we get $||Q(\sum_{i=1}^{\infty} c_i \xi_0^i)|| = \lambda' < \lambda$, where Q is the projection on $M(0)^{\perp} \cap H(\infty)$. Let Q_L be the projection on $H(\infty) \cap M_L(0)^{\perp}$. Then since $M_L(0) \uparrow M(0)$ we obtain $Q_L \downarrow Q$. So there exists $N_3 > 0$ such that

(4) $$||Q_n (\sum_{i=1}^{\infty} c_i\xi_0^i)|| < \lambda' + \varepsilon/4, \quad \text{for all } n > N_3 .$$

Let $N = \max(N_1,N_2,N_3)$ then by (4) we obtain that $||Q_N(\sum_{i=1}^{\infty} c_i\xi_0^i)|| < \lambda' + \varepsilon/4$. Hence we get

$$||Q_N(\sum_{i=N+1}^{\infty} c_i\xi_0^i) + Q_N(\sum_{i=1}^{N} c_i\xi_0^i)|| < \lambda' + \varepsilon/4.$$

So we get

$$||Q_N(\sum_{i=1}^{N} c_i\xi_0^i)|| - ||Q_N(\sum_{i=N+1}^{\infty} c_i\xi_0^i)||$$

$$\leq ||Q_N(\sum_{i=N+1}^{\infty} c_i\xi_0^i) + Q_N(\sum_{i=1}^{N} c_i\xi_0^i)|| \leq \lambda' + \varepsilon/4.$$

Hence

$$||Q_N(\sum_{i=1}^{N} c_i\xi_0^i)|| \leq \lambda' + \varepsilon/4 + ||Q_N(\sum_{i=N+1}^{\infty} c_i\xi_0^i)|| < \lambda' + \varepsilon/2.$$

Thus

$$||Q_N(\sum_{i=1}^{N} c_i\xi_0^i)||^2 < (\lambda' + \varepsilon/2)^2$$ and hence by (2) we get

$$||Q_N(\sum_{i=1}^{N} c_i\xi_0^i)||^2 < (\lambda' + \varepsilon/2)^2 + [(\lambda^2 - (\lambda' + \varepsilon/2)^2 -$$

$$\lambda^2 \sum_{i=N+1}^{\infty} |c_i|^2] < \lambda^2 - \lambda^2 \sum_{i=N+1}^{\infty} |c_i|^2 = \lambda^2 (\sum_{i=1}^{\infty} |c_i|^2 - \sum_{i=N+1}^{\infty} |c_i|^2).$$

Hence

$$||Q_N(\sum_{i=1}^{N} c_i \xi_0^i)||^2 < \lambda \sum_{i=1}^{N} |c_i|^2 \quad \text{and thus we get}$$

$$||\sum_{i=1}^{N} c_i \xi_0^i - (\sum_{i=1}^{N} c_i \xi_0^i \mid M_N(0))||^2 < \lambda^2 \sum_{i=1}^{N} |c_i|^2 ,$$

or

$$||\sum_{i=1}^{N} c_i(\xi_0^i - (\xi_0^i | M_N(0))||^2 < \lambda^2 \sum_{i=1}^{N} |c_i|^2 .$$

So we get $||\sum_{i=1}^{N} c_i \xi_{L,0}^i||^2 < \lambda^2 \sum_{i=1}^{N} |c_i|^2$, which implies

$$\sum_{i,j=1}^{N} c_i (\Sigma_N)_{ij} \overline{c_j} < \lambda^2 \sum_{i=1}^{N} |c_i|^2 .$$

Hence we get $\Sigma_N < \lambda^2 I_N$, which is a contradiction.

We now use this lemma to obtain some sufficient conditions for a process to be minimal full rank.

3.2 **Theorem.** Let ξ_n, $-\infty < n < \infty$ be a S.S.P. whose density f satisfies

$$0 < m(e^{i\theta})I \leq f(e^{i\theta}) \leq M(e^{i\theta}),$$

with $M(e^{i\theta})$ and $1/m(e^{i\theta})$ being summable. Then the process ξ_n, $-\infty < n < \infty$ is minimal full rank and we have

$$\sum = [\frac{1}{2\pi} \int_0^{2\pi} f^{-1}(e^{i\theta})d\theta]^{-1} .$$

Proof. From $m(e^{i\theta})I \leq f(e^{i\theta}) \leq M(e^{i\theta})I$ follows that $m(e^{i\theta})I_L \leq f_L(e^{i\theta}) \leq M(e^{i\theta})I_L$, hence $(1/M(e^{i\theta}))I_L \leq f_L^{-1}(e^{i\theta}) \leq (1/m(e^{i\theta}))I_L$. Thus

$$[\frac{1}{2\pi} \int_0^{2\pi} \frac{d\theta}{m(e^{i\theta})}]^{-1} I_L \leq [\frac{1}{2\pi} \int_0^{2\pi} f_L^{-1}(e^{i\theta})d\theta]^{-1}.$$

Now since $0 < \int_0^{2\pi} \frac{d\theta}{m(e^{i\theta})} < \infty$ choosing $\lambda = [\frac{1}{2\pi} \int_0^{2\pi} \frac{d\theta}{m(e^{i\theta})}]^{-1}$, by Lemma 3.1, we conclude that $\Sigma \geq \lambda I$. Hence the process is minimal full rank. Considering the innovation ζ_0 there exists $\Psi \in \bar{L}_2(f)$ such that $\zeta_0 = \bar{S}\Psi$. We note that, for any integer k, we have

$$(\zeta_0, \xi_k) = (\bar{S}\zeta_0 \,,\, \bar{S}\xi_0)_f = (\Psi, e^{-i\theta k})_f =$$
$$= \frac{1}{2\pi}\int_0^{2\pi} \Psi(e^{i\theta}) f(e^{i\theta}) e^{i\theta k} d\theta \,.$$

On the other hand, for any integer k, one can easily verify that

$$\frac{1}{2\pi}\int_0^{2\pi} e^{ik\theta}\, \Sigma d\theta = (\zeta_0, \xi_k).$$

Hence Ψf and Σ have the same Fourier coefficients. Thus $\Psi f = \Sigma$. Therefore by properties of $\bar{S}$ and $\bar{T}$ established earlier we have

$$\sum = (\zeta_0, \zeta_0) = (\bar{S}\Psi, \bar{S}\Psi) = (\Psi, \Psi)_f =$$
$$= \langle \bar{T}\Psi, \bar{T}\Psi \rangle_f = \frac{1}{2\pi}\int_0^{2\pi} \Sigma f^{-1}(e^{i\theta})\, \Sigma \, d\theta \,.$$

Hence one can verify that $\Sigma = \Sigma(\frac{1}{2\pi}\int_0^{2\pi} f^{-1}(e^{i\theta}) d\theta)\Sigma$ from which the result follows.

Lemma 3.1 may also be used to establish the following natural extension of Theorem 2.8 for the infinite dimensional stationary stochastic processes.

3.3 <u>Theorem</u>. Let ξ_n, $-\infty < n < \infty$ be a S.S.P. with spectral density f satisfying $0 < m(e^{i\theta})I \leq f(e^{i\theta}) \leq M(e^{i\theta})I$ a.e., where $M(e^{i\theta})$ is a summable scalar valued function. Let f_L be the top left $L \times L$ submatrix of f. Then the process ξ_n, $-\infty < n < \infty$ is minimal full rank if and only if there exists a constant μ such that

$$\int_0^{2\pi} (f_L(e^{i\theta}))^{-1}\, d\theta \leq \mu I_L \,,$$

uniformly for all L, $1 \leq L < \infty$. Furthermore

$$\sum = \lim_{L\to\infty} \left[\frac{1}{2\pi}\int_0^{2\pi} (f_L(e^{i\theta}))^{-1} d\theta \right]^{-1} .$$

Proof. The process ξ_n, $-\infty < n < \infty$ is minimal full rank if and only if $\Sigma \geq \lambda I$ for some $\lambda > 0$. But by Lemma 3.1, $\Sigma \geq \lambda I$ if and only if $\Sigma_L \geq \lambda I_L$ uniformly in L, $1 \leq L < \infty$. Hence the process ξ_n, $-\infty < n < \infty$, is minimal full rank if and only if $\Sigma_L \geq \lambda I_L$ uniformly in L, $1 \leq L < \infty$. Now suppose the process is minimal full rank, then $\Sigma_L \geq \lambda I_L$. Hence by Lemma 2.8, $\frac{1}{2\pi}\int_0^{2\pi}(f_L(e^{i\theta}))^{-1}\,d\theta = \Sigma_L^{-1} \leq \frac{1}{\lambda} I_L$. Now choose μ to be $\frac{1}{\lambda}$. To show the sufficiency suppose that $\int_0^{2\pi}(f_L(e^{i\theta}))^{-1}d\theta \leq \mu I_L$ uniformly in L. Then again by Theorem 2.8, $\Sigma_L^{-1} = \int_0^{2\pi}(f_L(e^{i\theta}))^{-1}d\theta \leq \mu I_L$ uniformly in L. Hence $\Sigma_L \geq (\frac{1}{\mu}) I_L$ uniformly in L, and $\Sigma \geq (\frac{1}{\mu})I$. Thus the process is minimal full rank.

Considering both sides of the relation

$$\Sigma_L = [\frac{1}{2\pi}\int_0^{2\pi}(f_L(e^{i\theta}))^{-1}d\theta]^{-1}$$

as infinite dimensional matrices with zeros in appropriate entries, and noting that then $\Sigma_L \to \Sigma$ elementwise we obtain

$$\Sigma = \lim_{L\to\infty}[\frac{1}{2\pi}\int_0^{2\pi}(f_L(e^{i\theta}))^{-1}d\theta]^{-1}.$$

3.4 **Examples.** One can easily construct spectral densities satisfying assumptions of Theorem 3.2 and 3.3. For instance in Theorem 3.2 we can take $f(\theta) = a(\theta)I$ with $a(\theta)$ being a scalar valued function such that both $a(\theta)$ and $\frac{1}{a(\theta)}$ are summable. For Theorem 3.3 one can take $f(\theta)$ to be the diagonal matrix-valued function whose nth diagonal element is $\theta^2 + \frac{1}{n}$. In this case $m(\theta)$ can be taken to be $m(\theta) = \theta^2$ for which $\frac{1}{m(\theta)} = \theta^{-2}$ is not summable on $[0,2\pi]$. However f does satisfy the condition of Theorem 3.3. Using Theorem 3.3, a simple calculation shows that Σ is the diagonal matrix whose nth diagonal element is given by $\frac{2\pi}{\sqrt{n}\,\arctan(2\sqrt{n}\,\pi)}$.

References

[1] R. Gandolli, Wide sense stationary sequences of distributions on Hilbert space and the factorization of operator valued functions. J.Math. Mech. 12 (1963), 893-910.

[2] A.N. Kolmogorov, Stationary sequences in Hilbert space.Bull. Math. Univ. Moscow, 2 (1941), 1-40.

[3] P. Masani, The prediction theory of multivariate stochastic processes, III. Acta Math. 104 (1960), 142-162.

[4] A.G. Miamee, An algorithm for determining the generating function and the linear prediction of an infinite dimensional stationary stochastic process. Sankhyā, Seriea A. 36 (1974).

[5] M.G. Nadkarni, Prediction theory of infinite variate weakly stationary stochastic processes. Sankhyā, Seriea A. 32 (1970), 145-172.

[6] R. Payen, Functions aléatoires du second order á valeurs dans un espaces de Hilbert. Ann. Inst.Henri Poincaré, 3 (1967), 323-396.

[7] H. Salehi, Application of the Hellinger integrals to q-variate stationary stochastic processes. Ark.Mat.7 (1967), 305-311.

[8] N. Wiener and P. Masani, The prediction theory of multivariate stochastic processes. I. Acta Math. 98 (1957), 111-150; II, Acta Math. 99 (1958), 93-137.

Arya-Mehr University of Technology,
Isfahan, Iran

Michigan State University, East
Lansing, Michigan 48824, USA

Michigan State University,
East Lansing, Michigan 48824,
USA

ON THE FACTORIZATION OF A NONNEGATIVE OPERATOR VALUED FUNCTION

A.G. Miamee and H. Salehi

Abstract

Let f be a nonnegative operator valued function on a Hilbert space. Suppose U is a unitary operator valued function. First the factorability of UfU^* is studied and some results are obtained. Using these results, a sufficient condition for the factorability of f in terms of its eigenvalues is given.

1. Introduction. Our main aim is to prove Theorem 3.4 which gives a sufficient condition for the factorability of a nonnegative operator valued function f in terms of its eigenvalues.

To accomplish our goal we will first study the factorability of UfU^*, where U is a measurable unitary operator valued function. (This problem was raised by M.G. Nadkarni in [7].) Here we obtain a result giving a sufficient condition under which the factorability of f and UfU^* are equivalent. This will make it possible to consider the factorization of UfU^* instead of f. In particular we will be interested in those function U's for which UfU^* is a diagonal operator.

In section 2 we set up necessary terminologies and state a known result. In section 3 we prove our main results.

2. Preliminaries. In this note all the Hilbert spaces are assumed to be separable. Let H and K be two Hilbert spaces. We denote by $B(H,K)$ the Banach space of all bounded linear operators on H into K and by $B^+(H,H)$ the class of all nonnegative operators in $B(H,H)$. Let f be a measurable $B(H,H)$-valued function on the unit circle. (In view of separability of Hilbert spaces involved, the notions of weak and strong measurability are equivalent. Hence the term measurability is used for either one). f is said to be weakly summable if for each x and y in H, $\int_0^{2\pi} |(f(e^{i\theta})x,y)|d\theta < \infty$, and uniformly summable if

$\int_0^{2\pi} ||f(e^{i\theta})|| d\theta < \infty$. We denote by $L_2(H)$ the Hilbert space of all measurable H-valued functions g for which $\int_0^{2\pi} ||g(e^{i\theta})||^2 d\theta < \infty$. The $L_2(H)$ inner product of two functions g_1 and g_2 is given by $(g_1,g_2) = \frac{1}{2\pi}\int_0^{2\pi}(g_1(e^{i\theta}), g_2(e^{i\theta}))d\theta$. A measurable $B(H,K)$-valued function $A = A(e^{i\theta})$ is said to be analytic if for each $h \in H$, Ah belongs to $L_2(K)$ and $\int_0^{2\pi} e^{-in\theta} A(e^{i\theta})h \, d\theta = 0$ for all $n < 0$.

2.1 Definition. Let f be a weakly summable $B^+(H,H)$-valued function on the unit circle. We say that f is factorable if there exists an analytic $B(H,H)$-valued function $\Phi = \Phi(e^{i\theta})$ on the unit circle such that

$$f(e^{i\theta}) = \Phi(e^{i\theta}) \Phi^*(e^{i\theta}) \quad \text{a.e.}$$

2.2. Remark. In definition 2.1 one can show that $\Phi(e^{i\theta})$ is actually of the form

$$\Phi(e^{i\theta}) = \sum_{k=0}^{+\infty} \Phi_k e^{ik\theta},$$

where Φ_k,s are constant bounded operators (see [6]).

In [9] Yu. A. Rozanov gave the following necessary and sufficient condition for the factorability of a weakle summable $B^+(H,H)$-valued function on the unit circle (see also [5]).

2.3 Theorem (Rozanov). Let f be a weakly summable $B^+(H,H)$-valued function. Then f is factorable if and only if there exists an analytic operator valued function Ψ such that

(a) $\Psi(e^{i\theta})H \subseteq f^{1/2}(e^{i\theta})H$,

(b) $\overline{f^{-1/2}(e^{i\theta}) \Psi(e^{i\theta})H} = \overline{f^{1/2}(e^{i\theta})H}$,

(c) $\int_0^{2\pi} ||f^{-1/2}(e^{i\theta})\Psi(e^{i\theta})h||^2 d\theta < \infty, \quad h \in H$,

where the symbol ―― denotes the closure operation and $f^{-1/2}(e^{i\theta})$ is the inverse operator from $f^{1/2}(e^{i\theta})H$ into $\overline{f^{1/2}(e^{i\theta})H}$.

3. Main Results. In this section we first prove a result on the factorability of UfU^*. Using this result we will obtain a sufficient condition for the factorability of f in terms of its eigenvalues (Theorem 3.4).

3.1 Theorem. Let $U = U(e^{i\Theta})$ be a measurable unitary $B(H,H)$-valued function. Suppose f is a uniformly summable $B^{+}(H,H)$-valued function such that $R(e^{i\Theta}) = \overline{f^{1/2}(e^{i\Theta})H}$ is a reducing subspace (closed) of $U(e^{i\Theta})H$. Suppose $\|f^{-1}(e^{i\Theta})\|\,\|f(e^{i\Theta})\|$ is summable ($f^{-1}(e^{i\Theta})$ is the inverse operator from $\overline{f(e^{i\Theta})H}$ into $\overline{f^{1/2}(e^{i\Theta})H}$). Then f is factorable if and only if UfU^* is factorable.

Proof. Suppose f is factorable as $f(e^{i\Theta}) = \Phi(e^{i\Theta})\Phi^*(e^{i\Theta})$, with $\Phi = \Phi(e^{i\Theta})$ being an analytic operator valued function. We will show that (a)-(c) of Theorem 2.3 hold for UfU^* with Ψ being Φ. By [1] we have

$$\overline{f^{1/2}(e^{i\Theta})H} = \overline{\Phi(e^{i\Theta})H}$$

Hence

$$\overline{\Phi H} = \overline{f^{1/2}H} = U(\overline{f^{1/2}(U^*H)}) = \overline{(Uf^{1/2}U^*)H} \quad \text{a.e.}$$

which is (a). Now by the definition of $f^{-1/2}f^{1/2}H = \overline{f^{1/2}H} = \overline{f^{1/2}H}$. Since $\overline{f^{1/2}H}$ is a reducing subspace for U we get

$$\overline{Uf^{1/2}U^*H} = \overline{Uf^{1/2}H} = U(\overline{f^{1/2}H}) = U(\overline{f^{-1/2}f^{1/2}H}) =$$

$$= \overline{Uf^{-1/2}f^{1/2}H} = \overline{Uf^{-1/2}(U^*f^{1/2}H)} = \overline{Uf^{-1/2}U^*f^{1/2}H} = \overline{Uf^{-1/2}U^*\Phi H}$$

which shows (b). Now by [1] there exists an operator valued function $C = C(e^{i\Theta})$ with norm one such that $\Phi(e^{i\Theta}) = f^{1/2}(e^{i\Theta})C(e^{i\Theta})$. So we have

$$\int_0^{2\pi} \|Uf^{-1/2}U^*\Phi\|^2 \, d\Theta \le \int_0^{2\pi} \|f^{-1/2}U^*\Phi\|^2 \le$$

$$\int_0^{2\pi} \|f^{-1/2}U^*f^{1/2}\| d\Theta \le \int_0^{2\pi} \|f^{-1/2}\|^2 \|f^{1/2}\|^2 d\Theta$$

$$= \int_0^{2\pi} \|f^{-1}\|\,\|f\| \, d\Theta < \infty .$$

This verifies (c) and hence completes the proof of one way. One can give a similar argument for the proof of the other way considering the fact that $Uf^{1/2}U^*H$ a reducing subspace for U^*.

3.2 <u>Lemma.</u> Let f be a uniformly summable $B^+(H,H)$-valued function which is factorable. Suppose that for almost all Θ's, $f(e^{i\Theta})$ in invertible and $\|f^{-1}(e^{i\Theta})\|\|f(e^{i\Theta})\|$ is summable. Let $l(e^{i\Theta})$ be an eigenvalue of $f(e^{i\Theta})$, which has a measurable eigenvector, say $v(e^{i\Theta})$. Then

$$\int_0^{2\pi} \log l(e^{i\Theta}) > -\infty.$$

<u>Proof.</u> Let $U = U(e^{i\Theta})$ be a measurable unitary valued function which maps $v(e^{i\Theta})$ to e_1 ($\{e_k\}_{k=1}^{\infty}$ is a complete orthonormal system for H). Then since $f^{1/2}H = H$ by Theorem 3.1 we have $UfU^* = \Phi\Phi^*$, with Φ being analytic. Hence

$$l(e^{i\Theta}) = (f(e^{i\Theta})v(e^{i\Theta}),\ v(e^{i\Theta})) = (f(e^{i\Theta})U^*(e^{i\Theta})e_1,$$

$$U^*(e^{i\Theta})e_1) = (U(e^{i\Theta})f(e^{i\Theta})U^*(e^{i\Theta}))e_1, e_1) = (\Phi^* e_1, \Phi^* e_1)$$

since $\Phi^* e_1 = \sum_{n=1}^{\infty} (\Phi^* e_1, e_n)$ we get $l(e^{i\Theta}) = \sum_{n=1}^{\infty} |(\Phi^* e_1, e_n)|^2$ (1)

But since $\Phi^*(e^{i\Theta}) = \sum_{k=0}^{\infty} e^{-ik\Theta}\, \Phi_k^*$ we see that

$$\Phi^*(e^{i\Theta})e_1,\ e_n) = \sum_{k=0}^{\infty} (\Phi_k^*\, e_1,\ e_n)e^{-ik\Theta}.$$

Now by the standard characterization for the factorability of a scalar valued nonnegative function we deduce that for some n, $\int_0^{2\pi} \log|(\Phi^*(e^{i\Theta})\, e_1,\ e_n)|^2\, d\Theta > -\infty$. Hence by (1) we get

$$\int_0^{2\pi} \log l(e^{i\Theta})d\Theta > -\infty.$$

3.3 <u>Lemma.</u> Let f be a measurable $B^+(H,H)$-valued function whose spectrum consists only of its eigenvalues. Suppose each eigenvalue has finite multiplicity. Let $l_1(e^{i\Theta}) \geq l_2(e^{i\Theta}), \ldots$ denote the eigenvalues of $f(e^{i\Theta})$ listed according to their multiplicities. Then l_i's are measurable and there exists a measurable unitary valued function such that

$$U(e^{i\Theta})f(e^{i\Theta})U^*(e^{i\Theta}) = \sum_{k=1}^{\infty} l_k(e^{i\Theta})\Omega_k(e^{i\Theta}),$$

where Ω_k's are constant one dimensional projections.

Proof. By a similar argument as in [2] p.653, one can show that $\sum_{k=1}^{q} l_k(e^{i\theta})$ is measurable for each $q \geq 1$. Hence it is clear that each $l_k(e^{i\theta})$ is measurable. Following the proof of [3] p.391 we can show that there exists a complete orthonormal sequence $\{u_k(e^{i\theta})\}_{k=1}^{\infty}$ consisting of eigenvectors of f which are measurable. Let $\{e_k\}_{k=1}^{\infty}$ be a complete orthonormal sequence in H. Define a unitary operator valued function U by setting $U(e^{i\theta})u_k(e^{i\theta}) = e_k$, $k \geq 1$. One can easily see that U has the desidered properties.

3.4 Main Theorem. Let f be a uniformly summable $B^+(H,H)$-valued function which is invertible for almost every θ on the unit circle and such that $||f^{-1}||\,||f||$ is summable.

(a) Suppose the spectrum of f consists only of eigenvalues, each one of which being of finite multiplicity. Then f is factorable if and only if for each k we have

$$\int_0^{2\pi} \log l_k(e^{i\theta})d\theta > -\infty,$$

where $l_1(e^{i\theta}) \geq l_2(e^{i\theta}) \geq l_3(e^{i\theta}) \geq \ldots$ are the eigenvalues of $f(e^{i\theta})$ listed according to their multiplicities.

(b) Suppose that

$$f(e^{i\theta}) = \sum_{k=1}^{\infty} \lambda_k(e^{i\theta})P_k(e^{i\theta}),$$

where P_k's are one-dimensional measurable projection valued functions. Assume that P_k's are mutually orthogonal. Then f is factorable if and only if for each k we have

$$\int_0^{2\pi} \log \lambda_k(e^{i\theta})d\theta > -\infty.$$

Proof. (a) If f is factorable then by Lemma 3.2 and Lemma 3.3 we see that (1) holds.

Now suppose that (1) holds and let $U = U(e^{i\theta})$ be the unitary operator valued function of Lemma 3.3. Then we have

$$U(e^{i\theta})f(e^{i\theta})U^*(e^{i\theta}) = \sum_{k=1}^{\infty} l_k(e^{i\theta})\,\Omega_k,$$

where for each k, Ω_k is the orthogonal projection on $\{e_k\}$.

By Theorem 3.1 f is factorable if and only if UfU^* is so. Now by the standard factorization result on scalar valued nonnegative functions there exist analytic functions $\varphi_k = \varphi_k(e^{i\theta})$ such that $l_k(e^{i\theta}) = |\varphi_k(e^{i\theta})|^2$. We observe that

$$||\sum_{k=1}^{N} \varphi_k(e^{i\theta})\Omega_k h||^2 = (\sum_{k=1}^{N} \varphi_k(e^{i\theta})\Omega_k h, \sum_{k=1}^{N} \varphi_k(e^{i\theta})\Omega_k h)$$

$$= (\sum_{k=1}^{N} l_k(e^{i\theta})\Omega_k h, h) \nearrow (\sum_{k=1}^{\infty} l_k(e^{i\theta})\Omega_k h, h).$$

Hence we have

$$||\sum_{k=1}^{N} \varphi_k(e^{i\theta})\Omega_k h||^2 \leq (U(e^{i\theta})f(e^{i\theta})U^*(e^{i\theta})h, h). \quad (2)$$

Similarly we have

$$||\sum_{k=N}^{n} \varphi_k(e^{i\theta})\Omega_k h||^2 = (\sum_{k=N}^{n} l_k(e^{i\theta})\Omega_k h, h) \quad (3)$$

By (2) and (3) it follows that $\sum_{k=1}^{\infty} \varphi_k(e^{i\theta})\Omega_k h$ converges.

Define an operator Φ from H into $L_2(H)$ by $\Phi h = \sum_{k=1}^{\infty} \varphi_k(e^{i\theta})\Omega_k h$. It is obvious from (2) that Φ is a bounded operator.

If we let $N \longrightarrow \infty$ in

$$||\sum_{k=1}^{N} \varphi_k(e^{i\theta})\Omega_k h||^2 = (\sum_{k=1}^{N} l_k(e^{i\theta})\Omega_k h, h) \quad (4)$$

we get

$$(UfU^*h, h) = ||\sum_{k=1}^{\infty} \varphi_k \Omega_k h||^2 = (\sum_{k=1}^{\infty} \overline{\varphi_k}\Omega_k h, \sum_{k=1}^{\infty} \overline{\varphi_k}\Omega_k h),$$

which means that $UfU^* = \Phi\Phi^*$. Now to complete the proof of (a) we must show that Φ is analytic, To see this we first note that for each positive integer m, $\Phi_m h = \sum_{k=1}^{m} \varphi_k(e^{i\theta})\Omega_k h$ has only nonnegative Fourier coefficients. We also observe that

$$\int_0^{2\pi} \|\sum_{k=N}^{n} \varphi_k(e^{i\theta})\Omega_k h\|^2 \, d\theta = \int_0^{2\pi} (\sum_{k=N}^{n} l_k(e^{i\theta})\Omega_k h, h) d\theta$$

$$= \int_0^{2\pi} (\sum_{k=1}^{n} l_k \Omega_k h, h) d\theta - \int_0^{2\pi} (\sum_{k=1}^{N} l_k \Omega_k h, h) d\theta \, .$$

Hence by the dominated convergence theorem we see that the sequence $\sum_{k=1}^{n} \varphi_k(e^{i\theta}) \, \Omega_k h$ is Cauchy in $L_2(H)$. Hence $\Phi(e^{i\theta})h$ cannot have negative Fourier coefficients. For each $n \geq 0$ we let $\Phi_n h$ be the nth Fourier coefficient of Φh. Then $\Phi h = \sum_{n=0}^{\infty} e^{in\theta} \Phi_n h$. Now $UfU^* = (\Phi^* h, \Phi^* h)$, together with the weak integrability of UfU^* and the closed graph theorem implies that Φ_n's are bounded. Hence UfU^* is factorable. Thus f is factorable.

(b) For each fixed positive integer k let $\varphi_k(e^{i\theta})$ be the unit vector in the range of $P_k(e^{i\theta})$. We claim that $u_k(e^{i\theta})$ is measurable. To see this let $\{e_n\}_{n=1}^{\infty}$ be a complete orthonormal system for H. Then for every positive integer n the function $P_k(e^i)e_n = (e_n, \varphi_k(e^i))\varphi_k(e^i)$ is measurable. Divide the unit circle as the disjoint union of a countable sequence $\{E_{nk}\}_{n=1}^{\infty}$ of measurable sets such that $P_k(e^{i\theta})e_n$ is different from zero on E_{nk} and zero on E_{mk}'s for all $m > n$. Then obviously we have

$$u_k(e^{i\theta}) = \frac{P_k(e^{i\theta})e_n}{\|P_k(e^{i\theta})e_n\|} \quad \text{if} \quad \theta \in E_{nk}.$$

Thus $u_k(e^{i\theta})$ is measurable. We further observe that $\lambda_k(e^{i\theta})$ is measurable, because

$$\lambda_k(e^{i\theta}) = (f(e^{i\theta})u_k(e^{i\theta}), \, u_k(e^{i\theta})) \, .$$

Now define the unitary operator valued function U by

$$U(e^{i\theta})u_k(e^{i\theta}) = e_k, \; k \geq 1.$$

This U is measurable. The rest of the proof of (b) is exactly the same as in part (a).

Theorem 3.4 is of the same type of a result due to R. Payen [8] p. 379.

3.5 Theorem. Let f be a $B^+(H,H)$-valued function which is uniformly summable. The UFU^* is factorable if and only if there exists a partial isometry valued function $V(e^{i\theta})$, with initial range in H and terminal range $U(e^{i\theta})f^{1/2}(e^{i\theta})U^*(e^{i\theta})H$ such that $U(e^{i\theta})f^{1/2}(e^{i\theta})U^*(e^{i\theta})V(e^{i\theta})$ is an analytic $B(H,H)$-valued function.

Proof. Sufficiency is clear. For necessity suppose that UfU^* is factorable as $UfU^* = \Phi\Phi^*$, where Φ is an analytic operator valued function. We then have

$$(U(f^{1/2}U^*)\ (Uf^{1/2}U^*) = \Phi\Phi^* .$$

Hence for each $h \in H$ we have

$$||(Uf^{1/2}U^*)\ h||^2 = ||\ \Phi^*h||^2 \tag{1}$$

Define $W(e^{i\theta})$ on $(Uf^{1/2}U^*)$ into Φ^*H by $W(e^{i\theta})((Uf^{1/2}U^*)\ h) = \Phi^*h$. Because of (1) we can extend this to an isometry W on $\overline{Uf^{1/2}U^*H}$ onto $\overline{\Phi^*H}$. We then have

$$W\ Uf^{1/2}U^* = \Phi^* \tag{2}$$

Taking adjoint on both sides of (2) we get $\Phi = Uf^{1/2}U^*W^*$. Letting W^* be V we get $Uf^{1/2}U^*V = \Phi$. Thus $Uf^{1/2}U^*V$ is an analytic operator valued function.

References

[1] R.G. Douglas, On majorization, factorization and range inclusion of operators on Hilbert space. Proc. Amer. Math.Soc. 17 (1966) 413-415.

[2] K. Fan, On a theorem of Weyl concerning eigenvalues of linear transformation I. Proc. Nat. Acad. U.S. 35 (1949) 652-655.

[3] W. Fieger, Die anwendung egiger mass-und integration theoretisch satze auf matrizelle Rieman steiltjies-integrale. Math. Ann. 150 (1963) 387-410.

[4] H. Helson, Lectures on Invariant Subspaces. Academic Press, New York, 1964.

[5] A.G. Miamee and H. Salehi, Necessary and sufficient condition for factorability of nonnegative operator valued functions on a Banach space. Proc. Amer. Math. Soc. 46 (1974) 43-50.

[6] A.G. Miamee and H. Salehi, Factorization of positive operator valued functions on a Banach space. Indiana Univ. Math. J. 24 (1974) 103-113.

[7] M.G. Nadkarni, Prediction theory of infinite variate weakly stationary stochastic processes. Sankhya Series A, 32, Part 2 (1970) 145-172.

[8] R. Payen, Founctions aleatoires du second order a valeurs dans un espace de Hilbert. Ann. Inst. Henri Poincare 3 (1967) 323-396.

[9] Yu.A. Rozanov, Some approximation problems in the theory of stationary processes. J. Multivariate Anal. 2 (1972) 135-144. Also the same volume, Erratum page 463.

[10] N. Wiener and P. Masani, The prediction theory of multivariate stochastic processes. I. Acta. Math. 98 (1957) 111-150.

[11] N. Wiener and P. Masani, On bivariate stationary processes and the factorization of matrix valued functions. Theor. Probability Appl., English Edition 4 (1959) 300-308.

Arya-Mehr University of Technology
Isfahan, I r a n

Michigan State University
East Lansing, Michigan 48824
U S A

[5] [illegible] Lectures on [illegible] Subspaces. [illegible] Press, New York, 19[illegible].

[6] G. [illegible] and [illegible], Necessary and sufficient conditions [illegible] [illegible] [illegible] [illegible] [illegible] Math. Soc. [illegible] (197[illegible]) [illegible].

[7] [illegible] and [illegible], [illegible] [illegible] of [illegible] [illegible] [illegible] [illegible] [illegible] Math. [illegible] [illegible] (19[illegible]) [illegible].

[8] [illegible] The [illegible] of [illegible] [illegible] [illegible] 42 [illegible] (197[illegible]).

[9] [illegible] [illegible] [illegible] [illegible] [illegible] [illegible] [illegible] 2[illegible]-[illegible].

[10] [illegible] [illegible] [illegible] approximation [illegible] [illegible] [illegible] [illegible] [illegible] [illegible].

[11] [illegible] Wiener [illegible] [illegible] [illegible] [illegible] [illegible] [illegible] [illegible] [illegible] [illegible] [illegible] (197[illegible]) [illegible].

[12] [illegible] Wiener and P. Masani, [illegible] [illegible] [illegible] [illegible] [illegible] and [illegible] [illegible] [illegible] [illegible] [illegible] [illegible] [illegible] [illegible].

Arya-Mehr University of Technology
Tehran, Iran

Michigan State University
East Lansing, Michigan 48824
U.S.A.

DISSIPATIVE RANDOM PROCESSES

W. Mlak

The present paper is partly expository. We shall be concerned with second order processes.

The motivation for the present paper goes back to results of Livsic and Jancevic [13], who introduced the term "dissipative" process. We generalize their definition and then are able to relate to dissipative processes suitable semi-groups of contractions. These semi-groups are just the analogons of non-stationary shifts - see Getoor[7] and Cramer [3]. Our processes are of operator type, as are the processes studied by Weron in [26] and [27].

The investigation of dissipative processes in [13] has been performed with the help of functional models of dissipative bounded operators - see [12]. The point was that the covariance of the process has been determined in terms of complete set of unitary invariants involved in the model. Our point of view is slightly different. We begin with covariance and then construct the related semi-group of contractions. Next we give some decomposition theorems. Later we discuss processes with dilatable semi-groups of contractions. We are then able to use Fourier analysis of contractions developed by Sz.-Nagy and Foiaş in [25] and by other authors. In particular, we can use related functional models of [25]. We do this for illustration in some simple situations using the so called concrete models due to Ahern and Clark [1] . These models apply to one-dimensional dissipative processes with discrete time.

Our approach is mostly of geometrical character. The related theorems are of qualitative type. The natural question arises: how covariance determines the spectral and other properties of involved contractions and their models ? We discuss some examples related to this question. Certainly, better results are available and one of the purposes of the present paper is just to show what still is to be done in the subject.

In all what follows we refer freely to [16] and [25] for dilation theory and to [6] and [9] for dilation free approach to models of some contraction operators.

1. Let $P = (\Omega, B, m)$ be a probability space. Suppose $x(t) \in L^2(P)$ for $t \geq 0$. Let $c(t,s) = E\, x(t)\overline{x(s)}$. We say that the process $x(t)$ is <u>dissipative</u>, if for every n, every n-tuple $t_1 \dots t_n \in \mathbb{R}_+ = (0,\infty)$ and $\lambda_1 \dots \lambda_n \in \mathbb{C}^1$ the function

$$\varphi(s) = E \,|\sum_{j=1}^{n} x(t_j + s)\lambda_j|^2 \quad , \quad (s \geq 0)$$

is a decreasing one. A standard example of dissipative process can be obtained as follows :

Let $T(t)$ $(t \geq 0)$ be a semi-group of contractions in $L^2(P)$ i.e. $||T(t)|| \leq 1$, $T(0) = I$ (identity operator) and $T(t+s) = T(t)T(s)$ for $t,s \geq 0$. Define $x(t) = T(t)x_0$ where $x_0 \in L^2(P)$. Then

$$\varphi(s) = \;||T(s)\sum_{j=1}^{n} x(t_j)\lambda_j||^2 = \sum_{j=1}^{n} T(s+t_j)\; x_0\lambda_j||^2$$

and because $T(s)$ are contractions, $\varphi(s)$ is a decreasing function.

In what follows we will show that a certain converse phenomen on takes place, namely, with every dissipative process $x(t)$ we can associate a semigroup of contractions $T(t)$ in the space of the process in such a way, that $x(t) = T(t)x(0)$ for $t \geq 0$. This semigroup is just the analogon of non-stationary shift considered by Getoor [7].

In general setting we replace the additive semigroup $\mathbb{R}_+$ of non-negative reals by an arbitrary semigroup and the scalar valued process by a vector one. Then the covariance becomes an operator valued function.

To begin with we denote by $L(H,K)$ the space of linear bounded operators on the Hilbert space H with values in the Hilbert space K. $L(H)$ stands for $L(H,H)$. We use usual notation for inner products norms, adjoints of operators etc. Let now Z be an arbitrary set.

<u>Definition 1.0</u> We say that the operator function $C(\cdot,\cdot) : Z \times Z \longrightarrow L(H)$ is <u>positive definite</u> if for every n, every $t_1 \dots t_n \in Z$, $f_1 \dots f_n \in H$ the inequality

$$\sum_{i,k} (C(t_i,t_k)f_i,f_k) \geq 0$$

holds true.

We write $C \gg 0$ if $C(\cdot,\cdot)$ is positive definite. It is easy to see that $C \gg 0$ if and only if the scalar function $c(\cdot,\cdot) : (Z \times H) \times (Z \times H) \longrightarrow \mathbb{C}$ defined as

$$c(\{t,f\}, \{s,g\}) = (C(t,s)f,g)$$

is positive definite in the usual sense. If this is the case, there is a gaussian process $y(t,f)$ with zero mean and such that

$$E\, y(t,f)\overline{y(s,g)} = (C(t,s)f,g)$$

for $t,s \in Z$; $f,g \in H$. Let H_c be the closed linear gaussian space spanned by this process. Then, as shown in [26] there is a family of operators $X(t) \in L(H,H_c)$ such that $y(t,f) = X(t)f$ for $t \in Z$, $f \in H$, and $C(t,s)$ can be factored as $C(t,s) = X(s)^* X(t)$, $t,s \in Z$. Consequently, $H_c = \bigvee_{t\in Z} X(t)H$ i.e. H_c is minimal. The minimality property determines $(X(\cdot),H_c)$ up to unitary equivalence, provided $C(t,s) = X(s)^* X(t)$. This means that if H'_c is a complex Hilbert space and $X_1(t) \in L(H,H'_c)$ and $C(t,s) = X_1(s)^* X_1(t)$ $(t,s \in Z)$ as well as $H'_c = \bigvee_{t\in Z} X_1(t)H$, then there is a unitary map $U : H_c \longrightarrow H'_c$ such that $UX(t) = X_1(t)$ for $t \in Z$. We see that, the minimality of H_c determines the expression (called later a canonical one) $C(t,s) = X(s)^* X(t)$ in a unique way modulo the unitary equivalence. Following the ideas developed by Weron in [26], [27] we define now a second order process over H and indexed by elements of Z, as a family of operators $X(t) \in L(H,K)$, $t \in Z$, where K is a Hilbert space. The covariance of $X(t)$ is defined as the operator function $C(t,s) = X(s)^* X(t)$ and the minimal space $\bigvee_{t\in Z} X(t)H$, the vector time domain of the process. It follows that every positive definite operator function $C(t,s) \in L(H)$ is a covariance of a suitable process $X(t) \in L(H,K)$ - see Weron [26]. Th. 4.6. If $K = H_c$ then $X(t)$ is called minimal.

<u>Definition 1.1</u> Let S be a multiplicative semigroup and $C(\cdot,\cdot) : S \times S \longrightarrow L(H)$ a positive definite function. We say that $C(\cdot,\cdot)$ is <u>dissipative</u> if for every n, every m-tuples $t_1 \dots t_n \in S$, $f_1 \dots f_n \in H$ and every $s \in S$ the inequality

$$\sum_{i,k} (C(st_i,st_k)f_i,f_k) \le \sum_{i,k} (C(t_i,t_k)f_i,f_k)$$

holds true.

The minimal process $X(t)$ with dissipative covariance is called a minimal dissipative process.

Definition 1.2 Let S be a semigroup and K a complex Hilbert space. The operator valued function $T(t) \in L(K)$ is called a semigroup over S, if $T(st) = T(s)T(t)$ $s,t \in S$. If $||T(t)|| \leq 1$ for $t \in S$, then $T(.)$ is called a semigroup of contractions over S. If e is a unit in S and $T(.)$ a semigroup over S, then $T(.)$ is called unital, if $T(e) = I$ (= identity operator in K).

Notice, that if S has unit e and $T(.)$ is a semigroup over S, then $T(e)$ in an idempotent. Hence, we can always suppose without any loss of generality that $T(.)$ is unital because $T(.)|\hat{H} = 0$, where $\hat{H} = (I - T(e))H$.

The next lemma is an operator analogon of a result of Getoor [7] - see also Cramer [3], Weron [26].

Lemma 1.0. Let S be a semigroup and let $X(t) \in L(H,H_c)$ be a minimal dissipative process indexed by S with covariance $C(t,s)$. Then there is a semigroup of contractions $T(t) \in L(H_c)$, $(t \in S)$ such that $C(ut,vs) = X(s)^*T(v)^*T(u)X(t)$ for $u,v,t,s \in S$.

Proof. Let $f_1 \dots f_n \in H$, $t_1 \dots t_n \in S$ and $s \in S$. Since C is dissipative then

$$|| \sum_j X(st_j)f_j||^2 = \sum_{i,k} (C(st_i,st_k)f_i,f_k) \leq$$

$$\leq \sum_{i,k} (C(t_i,t_k)f_i,f_k) = || \sum_j X(t_j)f_j ||^2$$

It follows that there is a unique contraction $T(s) \in L(H_c)$ such that $T(s)X(t)f = X(st)f$ for $s,t \in S$, $f \in H$. Since elements $X(t)f$ span H_c and $T(s_1s_2)X(t)f = X(s_1(s_2t))f = T(s_1)X(s_2t)f = T(s_1)T(s_2)X(t)f$, then $T(.)$ is a semigroup over S. We have for $f,g \in H$ $(T(u)X(t)f,T(v)X(s)f) = (X(ut)t,X(vs)g) = (X(vs)^* X(ut)f,g) = (C(ut,vs)f,g)$ which completes the proof.

Our basic theorem concerns unital S.

Theorem 1.0. Let S be semigroup with unit e and let $C(.,.) : S \times S \longrightarrow (H)$ be a dissipative positive definite operator function. Then there is a Hilbert space $\hat{H}$, an operator $R:H \longrightarrow \hat{H}$ and a unital semigroup $T(.)$ over S such that $C(t,s) = R^*T(s)^*T(t)R$ for $s,t \in S$.

<u>Proof.</u> Take simply $\hat{H} = H_c$, apply the previous lemma and define $R = X(e)$.

Notice that

$$(1.0) \qquad T(t)Rf = X(t.e)Rf = X(t)Rf \ ; \quad t \in S, \quad f \in H.$$

in the notation of the above theorem.

Theorem 1.0. is of dilation type. If S is a group, then dissipative covariance $C(t,s)$ is stationary i.e. $C(t,s) = \hat{C}(s^{-1}t)$ for some function $\hat{C} : S \longrightarrow L(H)$. $T(.)$ becomes then a unitary representation of S and the above theorem reduces to the theorem of Naimark*). In this way we get through the dilation theory the classical result, which states, that every stationary process over a Hilbert space H generates in a unique way a shift group $U(.)$ $(= T(.))$ in the space $H_c = \bigvee_{t \in S} U(t)H$ of the process, spanned by "translations" of the initial space H. Our theorem can be viewed as a generalization of this classical result and shows how dilations enter in an intrinsic way into the theory of second order process which are governed by some "laws of motion" - unitary ones in stationary case, and in a semigroup fashion for dissipative processes.

The next proposition shows that minimality of the vector time domain of dissipative process determines the related semigroup up to unitary equivalence. The proof of the proposition is a standard one in a dilation theory and will be omitted.

<u>Proposition 1.0.</u> Let S be a semigroup with unit. Assume that $T_i(t) \in L(K_i)$ $(i = 1,2)$ are unital semigroups over S. Suppose $R_i \in L(H,K_i)$ where H is a Hilbert space. If

$$(1.1) \qquad R^*_1 T_1(t)^* T_1(s) R_1 = R^*_2 T_2(t)^* T_2(s) R_2$$

for $t,s \in S$ and $K_i = \bigvee_{t \in S} T_i(t)R_iH$ then there is a unitary map $U : K_1 \longrightarrow K_2$ such that $UR_1 = R_2$ and $UT_1(t) = T_2(t)U$ for $t \in S$.

The above proposition says that the triple $(K,R,T(.))$, where $R : H \longrightarrow K = \bigvee_{t \in S} T(t)H$ and $T(.)$ is a semigroup over S, is determined uniquely by covariance up to unitary equivalence. We then call $T(.)$ minimal and the expression of dissipative covariance in the form $C(t,s) = R^* \, T(s)^* T(t) R$ the canonical one - it

*) see [16] and [26] for details.

coincides with the canonical expression defined before, because of (1.0).

<u>Corollary 1.0.</u> Suppose the assumption of th. 1.0. hold true and $C(e,e) = I_H$ (identity operator in H). Then R is an isometric embeding of H into H_c and when identifying Rf with $f \in H$, R^* is interpreted as the orthogonal projection P of H_c onto H. The assertion of Th. 1.0 reads then as follows :

(1.2) $\quad C(t,s)f = P\, T(s)^* T(t) f, \quad f \in H; \quad s,t \in S.$

It is then easy to see that in this case the unitary isomorphism of Prop. 1.0 keeps the vectors of H invariant. Notice, that also $H \subset H_c$, so the construction given in Th. 1.0 simply enlarges the space H.

<u>Corollary 1.1.</u> If $\dim H = 1$ in Th. 1.0 then H may be identified with $\mathbb{C}^1$ and $T(.)$ is cyclic namely $H_c = \bigvee_{t \in S} T(t)Rf_o$ where $f_o = 1$. In this case $C(.,.)$ is scalar valued. Conversely, if a given semigroup of contractions in K, say $V(t)$ $(t \in S)$ is cyclic i.e. $K = \bigvee_{t \in S} V(t)h_o$ for some vector h_o, then we can take the covariance $C_V(t,s) = (V(t)h_o, V(s)h_o)$. Then applying Th. 1.0 we get for minimal $T(.)$ $C_V(t,s) = (T(t)Rf_o, T(s)Rf_o)$ where $f_o = 1$ and establish unitary equivalence U which maps $V(t)h_o$ on $T(t)Rf_o$. It follows that there is a one-to-one correspondence between cyclic semigroups of contractions and dissipative scalar valued functions over S, via related minimal processes and their semigroup of contractions.

We consider next decomposition properties of dissipative processes. Suppose namely that S is a semigroup and $T(.)$: $S \longrightarrow L(K)$ a semigroup of contractions. Then, using standard Hilbert space operator methods one can prove that there is a unique decomposition $K = K_u \oplus K_c$ such that :

(i) K_u and K_c reduce all $T(s)$, $s \in S$.

(ii) The part $T_u(.)$ of $T(.)$ in K_u is a unitary representation of S.

(iii) There is no non-zero subspace of K_c which reduces $T(.)$ to a unitary representation of S.

The decomposition $T(.) = T_u(.) \oplus T_c(.)$ is called the canonical decomposition, $T_u(.)$ the unitary part of $T(.)$, $T_c(.)$ the completely non-unitary (shortly c.n.u.) part of $T(\cdot)$. If $T(\cdot) = T_c(\cdot)$ then we say that $T(\cdot)$ is c.n.u.

The canonical decomposition when applied to semigroups of dissipative processes suggests the following definitions.

Definition 1.3. Let $T(\cdot)$ be the minimal semigroup of contractions of the dissipative process $X(t)$. We say that $X(t)$ is semi-stationary if $T(\cdot)$ is a unitary representation of S. $X(\cdot)$ is called purely dissipative if $T(\cdot)$ is a completely non-unitary semigroup of contractions over S.

Definition 1.4. Let $X(t) \in L(H,K_X)$, $Y(t) \in L(H,K_Y)$ be two processes indexed by t Z. $V(\cdot) = X(\cdot) \oplus Y(\cdot)$ for the process $V(t) \in L(H,K_X \oplus K_Y)$ defined as $V(t)f = X(t)f \oplus Y(t)f$ for $t \in Z$, $f \in H$. If $V(\cdot) = X(\cdot) \oplus Y(\cdot)$ then we say that $X(\cdot)$ and $Y(\cdot)$ are uncorrelated components of $V(\cdot)$.

Now we are able to prove the following theorem.

Theorem 1.1. Let $C(\cdot,\cdot) : S \times S \longrightarrow L(H)$ be the covariance of the minimal dissipative process $X(t) \in L(H,H_c)$. We assume that S is unital. Let $T(\cdot)$ be the minimal semigroup of contractions over S related to $X(t)$. Then $X(\cdot) = X_u(\cdot) \oplus X_c(\cdot)$ where $X_u(\cdot)$ is a semi-stationary process and $X_c(\cdot)$ a purely dissipative one.

Proof. Let $T(\cdot) = T_u(\cdot) \oplus T_c(\cdot)$ be the canonical decomposition of $T(\cdot)$ and $C(t,s) = R^*T(s)^*T(t)R$ that one of $C(\cdot,\cdot)$. Let P_u, P_c be the orthogonal projections on the unitary and c.n.u. part of H_c respectively. Then the desired processes are defined as

$$X_u(t)f = T_u(t)P_uRf$$

$$X_c(t)f = T_c(t)P_cRf$$

for $t \in S$, $f \in H$.

Notice that vector time domain of $X_u(\cdot)$ ($X_c(\cdot)$ respectively) equals to P_uH_c (P_cH_c respectively) and $T(\cdot)$ and $T_c(\cdot)$ are minimal semigroups related to $X_u(\cdot)$ and $X_c(\cdot)$ respectively.

Corollary 1.2. If S is a subsemigroup of the group G and S contains the unit of G, then $T_u(\cdot)$ can be extended to a unitary representation $U(\cdot) : G_s \longrightarrow L(P_uH_c)$ of the group generated by S. Then the related process $X_u(\cdot)$, when extended as a function of variable runnig over G_s becomes stationary.

We will discuss briefly a special kind of processes. Suppose namely that S is an abelian semigroup with additivelly written semigroup operation and unit denoted by 0.

Definition 1.5. The positive definite function $C(\cdot,\cdot) : S \times S \longrightarrow L(H)$ is called conservative if $C(t+n,s+n) = C(t,s)$ for $s,t,n \in S$.

It is easy to see that conservative $C(\cdot,\cdot)$ is dissipative. The related process $X(\cdot)$ will be then called conservative and the corresponding semigroup $T(t) \in L(H_c)$ will be an isometric representation of S. Assuming additionally that $C(0,0) = I$ we will have $H \subset H_c$ and $X(t)f = T(t)f$ for $t \in S$, $f \in H$. By Ito theorem [10] there is a space $K \supset H_c$ and a unitary representation $U(\cdot) : S \longrightarrow L(K)$ such that $T(t)h = U(t)h$ for $h \in H_c$. Now, if S is a subsemigroup of the abelian group G and S generates algebraically G, then $U(\cdot)$ has a unique extension to the unitary representation $\hat{U}(\cdot) : G \longrightarrow L(K')$ where $K' = \bigvee_{t \in G} U(t)H_c$. If follows that the covariance $C(t,s)$ can be extended to a covariance $\hat{C}$ of a stationary process by formula $(\hat{C}(t,s)f,g) = (\hat{U}(t)f,\hat{U}(s)g)$ for $f,g \in H_c$.

Notice that a conservative process need not be semi-stationary, simply because there are semigroups of isometries which are not unitary representations. In general, every semigroup of contractions $T(\cdot)$ can be decomposed as $T(\cdot) = T_i(\cdot) \oplus T_c(\cdot)$ where $T_i(\cdot)$ is a semigroup of isometries and no non-zero part of $T_c(\cdot)$ is a semigroup of isometries. Certainly $T_c(\cdot)$ will correspond to purely dissipative process, but $T_i(\cdot)$ can correspond to such one as well. The example below will illustrate general ideas.

Example. Let f_n, $n = 0,1,2,\dots$ be a sequence of second order random variables. Let $S = \mathbb{Z}^+$ - the additive semigroup of non-negative integers ans suppose that the related covariance of the proces $\{f_n\}$ is conservative i.e.

$$C(m,n) = E\, f_m \bar{f}_n = (f_m,f_n) = (f_{m+1},f_{n+1})$$

for all $m,n \in \mathbb{Z}^+$. It follows from Th. 1.0. that there is a unique isometry V in H_c such that $(f_m,f_n) = (V^m f_0, V^n f_0)$, provided $E|f_0|^2 = 1$. V is cyclic that is $H_c = \bigvee_{n \geq 0} V^n f_0$. Since

$V^*V = I$ then $(V^m f_0, V^n f_0) = C(m-n,0)$ if $m \geq n$ and $(V^m f_0, V^n f_0) = C(0,n-m)$ if $m \leq n$. Hence $\{C(m,n)\}$ is a Toeplitz matrix. Now, if $\rho = \inf_{p(0)=0} ||f_0 - p(V)f_0||^2 = 0$ where p ranges over analytic polynomials, then V is unitary. Indeed, then there exists a sequence

$h_n = \sum_{k|1}^{n} a_{k,n} V^k f_0$ such that $h_n \longrightarrow f_0$. Hence $V^*f_0 =$

$\lim \sum_{k|1}^{n} a_{k,n} V^{k-1} f_0$ and consequently $VV^*f_0 = f_0$. It follows that the range of V is all H_c, because f_0 is cyclic for V; hence V is unitary and $H_c = H_c^{(u)}$. If V is unitary then $VV^*f_0 =$

$= VV^{-1}f_0 = f_0$ and then $\rho = 0$ because $V^*f_0 = \lim g_n$ where

$g_n = \sum_{k|0}^{n} b_{k,n} V^k f_0$. The expression $||f_0 - p(V)f_0||^2$ can be computed in terms of entries of covariance matrix and coefficients of p. To do this let U be a unitary extension of V, and E the spectral measure of U. Define the scalar measure $\mu(\sigma) = (E(\sigma)f_0,f_0)$ for Borel subsets of the unit circle Γ. We define $V^{(n)} = V^n$ for $n \geq 0$ and $V^{(n)} = V^{*n}$ for $n \leq 0$. Then defining $c_n = (V^{(n)}f_0,f_0)$ we have $c_n = (U^n f_0,f_0) = (V^{(n)}f_0,f_0) = \int_\Gamma z^n d\mu$. We take the polynomial $p(z) = -\sum_{k|1}^{n} \lambda_k z^k$ and putting $\lambda_0 \equiv 1$ get that

$||f_0 - p(V)f_0||^2 = \sum_{i,k|0}^{n} c(i-k)\lambda_i \overline{\lambda}_k = \varphi(\lambda_1 \dots \lambda_n)$. Now $\rho > 0$ if and only for some $\eta > 0$, for all $n \geq 1$, $\eta \leq \min_{\lambda_1 \dots \lambda_n} \varphi(\lambda_1 \dots \lambda_n)$.

In fact one can prove that $\rho = \lim \rho_n$ where $\rho_n = \min_{\lambda_1 \dots \lambda_n} \varphi(\lambda_1 \dots \lambda_n)$.

In this may we arrive to a criterium for V to be not unitary in terms of covariance. If V has no unitary part then it has no semi-stationary one.

2. The example at the end of the previous section shows that the covariance $C(m,n)$ of a dissipative process can be a function merely of $(m-n)$ but the process will be not semi-stationary. A similar phenomena appears for a wide class of dissipative processes. To be more precise, such processes can have

subordinate ones with covariance of conservative character. For processes over the semigroup $S = \mathbb{R}_+$ the corresponding conservative parts appear on a purely analytic way in [11] and [13] and within some geometrical frames for processes over $S = \mathbb{Z}^+$ and $S = \mathbb{R}_+$ in [16] and in [25], in connection of triangulation of contractions and Wold type decompositions of dilation spaces. The given proofs however in this situation depended essentially on the order structure of related semigroups as well on special properties of dilation spaces. In what follows we will show that similar effects appear in more general circumstances, namely for process over semigroups which semiorder suitable groups. Our arguments will not appeal to special structure properties of dilation spaces as well to their minimality. They are of prediction theoretical character and work pretty well for what traditionally could be called the vector time domain analysis of random fields rather than processes, when the "time" is ordered in an archimedean way.

In all what follows G is an abelian group with group operation written additivelly and neutral element 0. We assume once for all that

(2.0) The subsemigroup $S \subset G$ contains 0 and generates algebraically all of G.

<u>Definition 2.0.</u> Let $U(.) : G \longrightarrow L(K)$ be a unitary representation of G and let $T(.) : S \longrightarrow L(H)$ be a unital semigroup of contractions. We say that $U(.)$ is a unitary dilation of $T(.)$ if $\hat{H} \subset K$ and $T(t)f = P\,U(t)f$ for $t \in S$, $f \in \hat{H}$, where P is the orthogonal projection of K on $\hat{H}$.

The next lemma generalizes the theorem Th. 3.2. I of [25] and is a refinement of Th. 5.1 of [16].

<u>Lemma 2.0.</u> Let $U(.)$ be a unitary dilation of the unital semigroup of contractions $T(.) : S \longrightarrow L(\hat{H})$. Suppose that $\hat{H} = H' \oplus H''$ is the canonical decomposition such that the part $T'(.)$ of $T(.)$ in H' is a unitary representation of S and $T''(.)$, the part of $T(.)$ in H'' is a c.n.u. semigroup of contractions. Then $H' = \bigcap_{u \in G} U(u)\hat{H}$.

<u>Proof.</u> Notice first that since $T(u)f = PU(u)f$ for $u \in S$ then $T(u)^*f = PU(-u)f$ for $u \in S$. Suppose now that $f \in H'$. We will prove that for $u_1 \ldots u_n \in S$ $U(-S)\ U(u_1 + \ldots + u_n)\, f \in H'$.

The proof will be by induction. If $u \in S$ then, since $f \in H'$ we have $||T(u)f||=||f||$, $||T(u)f||=||PU(u)f||$. Since $||f||=||U(u)f||$ then $T(u)f = U(u)f \in H' \subset \hat{H}$. If $u \in (-S)$ then $||T(u)^*f||= = ||f|| = ||U(-u)f||=||PU(-u)f||$ which proves that $T(u)^*f = = U(-u)f \in H'$. Suppose now that $g = U(u_1+\dots+u_n)f = T_{u_n}T_{u_{n-1}} \dots T_{u_1} f \in H'$ for $f \in H'$ and any $u_i \in SU(-S)$, where $T_u = T(u)$ if $u \in S$ and $T_u = T(-u)^*$ if $u \in (-S)$. We take $u_{n+1} \in SU(-S)$. Then, since $g \in H'$ then $||T_{u_{n+1}} g||=||g||= ||PU(u_{n+1})g|| = = ||U(u_{n+1})g||$ which implies that $T_{u_{n+1}} g = U(u_{n+1})g \in H'$ i.e. $T_{u_{n+1}} T_{u_n} \dots T_{u_1} f = U(u_1+\dots+u_{n+1})f \in H'$. This completes the proof of the inclusion $H' \subset \bigcap_{u\in G} U(u)H' \subset \bigcap_{u\in G} U(u)\hat{H}$, because S generates G. Now, if $f \in \bigcap_{u\in G} U(u)\hat{H}$ then $U(u)f \in H$ for $u \in G$ and consequently $T_u f = U(u)f$ for $u \in S \cup (-S)$. Since $\bigcap_{u\in G} U(u)\hat{H}$ is invariant under $U(.)$, it is invariant under T_u for $u \in S \cup (-S)$, which by the last equality means that $\bigcap_{u\in G} U(u)\hat{H}$ reduces $T(.)$ to a unitary representation of S. Hence $\bigcap_{u\in G} U(u)\hat{H} \subset H'$ q.e.d.

Let $U(.) : G \longrightarrow L(K)$ be a unitary dilation of $T(.) : S \longrightarrow L(\hat{H})$ We define present and past spaces

$$M_+ = \bigvee_{t\in S} U(t)\hat{H}, \qquad M_- = \bigvee_{-t\in S} U(t)\hat{H}$$

and the corresponding remote pasts

$$R_+ = \bigcap_{t\in S} U(t)M_+, \qquad R_- = \bigcap_{-t\in S} U(t)M_-$$

Since G is abelian, R_+ and R_- reduce $U(.)$ to unitary representation because S generates G.

<u>Lemma 2.1.</u> Let $T(.) : S \longrightarrow L(\hat{H})$ be a unital semigroup of contrations and $U(.) : G \longrightarrow L(K)$ its unitary dilation i.e. $T(t)f = PU(t)f$ for $t \in S$, $f \in H$. Then :

(i) If $h \in U(t)M_+$, $t \in S$ then $Ph = T(t)PU(-t)h$.

(ii) If $h \in U(-t)M_-$, $t \in S$, then $Ph = T(t)^* PU(t)h$.

(iii) If $h \in R_+$ (R_- respectively) then $PU(t)h = T(t)Ph$ ($PU(-t)h= T(t)^* Ph$) for $t \in S$.

<u>Proof.</u> To prove (i) we take $h = U(t)U(s)f$ where $t,s \in S$ and $f \in H$. Then $Ph = PU(t)U(s)f = PU(t+s)f = T(t)T(s)f = T(t)PU(-t)U(t+s)f = T(t)PU(-t)h$. Since h of the above form span $U(t)M_+$, the proof is complete. Relation (ii) follows by symmetry.

To prove (iii) notice that R_+ reduces $U(\cdot)$. So, if $h \in R_+$ and $t \in S$ then $h = U(-t)g$ with some $g \in R_+$. Since $g \in R_+ \subset U(t)M_+$ then by (i) $Pg = T(t)PU(-t)g$ which implies that $PU(t)h = T(t)Ph$.

The above lemma generalizes results of [16] and of [17]. We derive now several consequences of this Lemma.

<u>Proposition 2.0.</u> Suppose the assumptions of Lemma 2.1 hold true. Let P_- (P_+) be the orthogonal projection (within K) onto R_- (R_+ respectively). Then

$$(2.1) \qquad P_- T(u)f = U(u)P_-f \quad (P_+T(u)^*f = U(u)P_+f) \quad \text{for} \quad u \in S \text{ and } f \in \hat{H}.$$

<u>Proof.</u> Let $f \in \hat{H}$, $u \in S$ and $g \in K$. Since R_- reduces $U(\cdot)$, there is $g_u \in R_+ \subset U(u)M_+$, such that $P_-g = U(u)g_u$, $u \in S$. It follows from (ii) of Lemma 2.0 that $T(u)^*PU(u)g_u = Pg_u$. Hence $(P_-T(u)f,g) = (f,T(u)^*PP_-g) = (f,T(u)^*PU(u)g_u) = (f,Pg_u) = (f,g_u) = (f,U(-u)P_-g) = (U(u)f,P_-g) = (P_-U(u)f,g)$ which implies $P_-T(u)f = P_-U(u)P_-f$ because R_- reduces $U(u)$.

The above proposition has been proved for the case $S = \mathbb{Z}^+$ in [25], as a consequence of some properties of asymptotic behaviour of contraction semigroup in question and structure properties of the minimal dilation space.

The following proposition generalizes results proved in [16] and [17].

<u>Proposition 2.1.</u> Suppose that the assumptions of Lemma 2.1 hold true. Then the unitary part H' of canonical decomposition related to $T(\cdot)$ satisfies the equality

$$H' = R_+ \cap R_-$$

<u>Proof.</u> Suppose $h \in R_+ \cap R_-$. Then by (iii) of Lemma 2.1.

$$PU(t)h = T(t)PH,$$

for $t \in S$. It follows that

$$u(t)(I - P)h \perp H \quad \text{for} \quad t \in S.$$

Hence

$$(I - P)h \perp M_-$$

But $(I - P)h = h - Ph \in M_-$ because $h \in M_-$ and $Ph \in M_-$. It follows that $h = Ph$ i.e. $R_+ \cap M_- \subset H$. Consequently $R_+ \cap R_- \subset H$.

Since $R_+ \cap R_-$ reduces $U(\cdot)$, $R_+ \cup R_- \subset \bigcap_{u \in G} U(u)\hat{H} = H'$ by Lemma 2.0. Now, if $h \in H'$ then for $t,s \in S$ $h = U(t)U(s)g_{t,s}$ with suitable $g_{t,s} \in H$ by Lemma 2.0 again. Since $U(s)g_{t,s} \in M_+$ and t is arbitrary we conclude that $h \in R_+$. By similar token $h \in R_-$ which proves the claim. To complete the series of preparatory propositions concerning unitary dilations of semigroups of contractions we will prove a generalization of [25] 3.1.II. Our proof does not appeal to structure of dilation spaces as well to its minimality.

Proposition 2.2. Let G be the additive subgroup of the additive group of reals and define $S = \{t \in G : t \geq 0\}$. Suppose $T(t)f = PU(t)f$ $t \in S$, $f \in \hat{H}$ where $T(t)$ is unital semigroup of contractions : $T(\cdot) : S \longrightarrow L(\hat{H})$, and $U(\cdot)$ its unitary dilation : $U(\cdot) : S \longrightarrow L(K)$, $K \supset \hat{H}$, P = orthogonal projection on $\hat{H}$. Then for $f \in \hat{H}$

(2.2) $\lim_{t \to \infty} U(t)T(t)^*f = P_+f$

(2.3) $\lim_{t \to \infty} U(t)^* T(t)f = P_-f$

where P_+ and P_- are projections on suitable remote pasts formed with the help of $U(\cdot)$ (which need not be minimal).

Proof. We will prove the first relation. The other one follows by symmetry.

Since for $s \leq t$ $a(t,s) = ||U(t)T(t)^*f - U(s)T(s)^*f||^2 = ||T(s)^*f||^2 - ||T(t)^*f||^2$ and $T(u)$ are contractions then $\lim_{t,s \to \infty} a(t,s) = 0$ and consequently $\lim_{t \to \infty} U(t)T(t)^*f = \hat{f}$ exists. Let P_t be the projection on $U(t)M_+$. Then $P_t \longrightarrow P_+$ strongly. By Lemma 2.1 (i) for every $h \in K$ we have

$$(T(t)PU(-t)P_t h,f) = (PP_t h,f)$$

and consequently

$$(P_t h,f) = (P_t h,U(t)T(t)^*f)$$

which by limit passage proves that

$$(P_+h,f) = (P_+h,\hat{f}) \quad \text{for } h \in K.$$

It follows then that $P_+f = P_+\hat{f}$ and consequently $||\hat{f}-P_+f||^2 = ||\hat{f}||^2 - ||P_+f||^2$. Notice now that $||\hat{f}||^2 = \lim||T(t)^*f||^2$. On the other hand by Lemma 2.1. (i)

$$||T(t)^*f||^2 = (PU(-t)P_tf, T(t)^*f) +$$
$$+ (PU(-t)(I-P_t)f, T(t)^*f) = ||P_tf||^2$$

because $U(t)T(t)^*f \in U(t)M_+$. By limit passage we conclude therefore that $||\hat{f}||^2 = ||P_+f||^2$ which completes the proof of (2.2). Relation (2.3) follows by symmetry.

Remark 2.0. It follows from the above proposition that

$$\lim T(t)T(t)^*f = PP_+f, \quad \lim T(t)^*T(t)f = PP_-f$$

In case of $S = \mathbb{Z}^+$ or $\mathbb{R}^+$ these relations have been proved in [25], and in an analytic way for $T(t)$, $(t \geq 0)$ with bounded generator, in [11].

We will deal now with unitary dilatable semigroup of contractions and restrict our considerations to processes with such semigroups. We are then able to prove some characterizations of parts of the processes in terms of Wold type decompositions of some dilation spaces.
We do this using previous results of the present section.

Theorem 2.0. Let $X(t) \in L(H,H_c)$ be the minimal dissipative process over the semigroup S which satisfies (2.0). Let $C(\cdot,\cdot)$ be the covariance of $X(\cdot)$ and suppose $C(0,0) = I$. Denote by $T(\cdot)$ the semigroup of contractions in H_c corresponding to $X(\cdot)$, and suppose that $U(\cdot) : S \longrightarrow L(K)$, $K \supset H_c$ is a unitary dilation of $T(\cdot)$. Let P_- be the orthogonal projection onto the remote past $R_- = \bigcap_{-t\in S} U(t)M_-$ where $M_- = \bigvee_{-t\in S} U(t)H_c$. Then $X(\cdot) = X'(\cdot) \oplus X''(\cdot)$ where for $t \in S$, $f \in H$

$$X'(t)f = P_-T(t)f,$$
$$X''(t) = (I_K - P_-)T(t)f.$$

$X'(\cdot)$ is conservative and vanishes if and only if $P_-f = 0$ for every $f \in H$. Moreover, $X''(\cdot)$ vanishes if and only if $X(\cdot)$ is conservative.

Proof. That $X(\cdot) = X'(\cdot) \oplus X''(\cdot)$ follows immediately from definitions of $X'(\cdot)$ and $X''(\cdot)$. Next, the covariance of $X'(\cdot)$ equals to $(P_-T(t)f, P_-T(s)f) = (U(t-s)P_-f, P_-f)$ by (2.1) of Proposition 2.0. Hence $X'(\cdot)$ is conservative. Since $X'(t)f = P_-T(t)f = U(t)P_-f$ then $X'(\cdot)$ vanishes if and only if $P_-f = 0$ for all $f \in H$.

To prove the last statement it is enough to prove that $X'(.) = X(.)$ if and only if $X(.)$ is conservative. Since $X'(.)$ is conservative, it is sufficient to show that if $X(.)$ is conservative, then $X(.) = X'(.)$. But if $X(.)$ is conservative then $T(.)$ is a semigroup of isometries. Hence $||T(t)f||=||f||= ||U(t)f||=||PU(t)f||$ for $t \in S$ and $f \in H$. It follows that $U(t)f = T(t)f$ and consequently $U(t)f \in H_c$ for $t \in S$ and $f \in H$. We derive therefore that

$$f \in U(-t)U(-s)H_c \subset U(-t)M_-$$

for all t. Hence $f \in R_-$ which implies that $X(t)f= T(t)f = U(t)f= U(T)P_-f = P_-T(t)f = X'(t)f$ for $t \in S$, $f \in H$ q.e.d.

<u>Corollary 2.0.</u> Notice that $P_-f = 0$ for $f \in H$ if and only if $R_- = \{0\}$. Obviously if $R_- = \{0\}$ then $P_-f = 0$ for $f \in H$. Suppose that $P_-f = 0$ for every $f \in H$. Then

$$P_-T(s)f = U(s)P_-f = 0$$

for $s \in S$ and consequently, since P_- commutes with $U(.)$ then for $t,s \in S$

$$U(-t)P_-T(s)f = P_-U(-t)T(s)f = 0$$

But H_c is spanned by elements $T(s)f$ where $s \in S$ and $f \in H$. It follows that $P_-M_- = \{0\}$ i.e. $R_- = \{0\}$.

<u>Corollary 2.1.</u> Suppose in Th. 2.0 $S = \{t \geq 0;\ t \in G\}$ where G is an additive subgroup of reals. Then the semigroup $T(.)$ related to $X(.)$ is unitarily dilatable (see (2.5) below). We claim that $P_-f = 0$ for a given $f \in H$ if and only if $\lim T(t)^*T(t)f=0$ or equivalently in terms of covariance, if and only if $\lim_{t\to\infty} (C(t,t)f,f) = 0$. To prove this let Q_t be the projection on $U(-t)M_-$ $(t \in S)$. Exactly as in the proof of Proposition 2.2., when using (ii) of Lemma 2.1 we get that for every $f \in H$ $||T(t)f||^2 = (C(t,t)f,f) = ||Q_tf||^2$. Since $Q_t \longrightarrow P_-$ strongly, the assertion follows immediately.

<u>Corollary 2.2.</u> Under the assumptions of Th 2.0 $X'(t)f \perp X''(s)g$ for $t,s \in S$ and $f,g \in H$. It follows that the covariance C is a sum of covariance of $X'(.)$ and $X''(.)$, more precisely

$$(C(t,s)f,g) = (X'(t)f,X'(s)g) + X''(t)f,X''(s)g)$$

Since $C(.,.)$ is dissipative and $X'(.)$ is conservative then $X''(.)$ is dissipative. Consequently, the canonical expression of its covariance is of the form

$$C''(t,s) = R^*\hat{T}(s)^*\hat{T}(t)R$$

where $R : H \longrightarrow H_{C''}$. Both $H_{C''}$ and $\hat{T}(\cdot)$ are minimal. It follows then that

$$||\hat{T}(t)Rf||^2 = ||X''(t)f||^2 = ||(I_K - P_-)T(t)f||^2$$

which for $t = 0$ implies that $||Rf|| = ||(I_K - P_-)f||^2$. We thus see that $X''(\cdot)$ can be viewed as equivalent to the process $Y(t)h = \hat{T}(t)h$ where $h \in \overline{RH}$ .The closure $\overline{RH}$ is mapped in a unitary way on $\overline{(I_K - P_-)H}$ by continuous extension of the map $V: RH \longrightarrow (I_K - P_-)H$ defined as $VRf = (I_K - P_-)f$. The covariance C_Y of $Y(t) \in L(\tilde{H},\tilde{H}_{C_Y})$ where $\tilde{H}_{C''} = \tilde{H}_{C_Y} = \bigvee_{t \in S} \tilde{T}(t)\tilde{H}$ will now satisfy the equality $C_Y(0,0) = I_{\tilde{H}}$. Suppose for simplicity that S is such as in Cor. 2.1. Then for $f \in H$

$$||\tilde{T}(t)Rf|| = ||T(t)f - P_-T(t)f|| = ||T(t)f - U(t)P_-f|| =$$

$$||U(-t)T(t)f - P_-f|| \underset{t\to\infty}{\longrightarrow} 0$$

by (2.3) of Prop. 2.2. It follows that $\tilde{T}(t)h \underset{t\to\infty}{\longrightarrow} 0$ for every $h \in \tilde{H}$ and consequently $\tilde{T}(t)h \underset{t\to\infty}{\longrightarrow} 0$ for every $h \in \tilde{H}_{C_Y}$. Hence, if we are interested in processes over $S \subset \mathbb{R}_+$ without conservative part, especially, if we are interested in analytical properties of covariances, then we have to deal merely with such processes, whose covariance $C(t,t) \underset{t\to\infty}{\longrightarrow} 0$ weakly. The reason is that if $X(\cdot)$ is conservative over $S \subset \mathbb{R}_+$, S as in Cor.2.1, then

$$(2.4) \qquad (X(t)f, X(s)g) = \int_{\hat{G}} \langle t - s, \lambda \rangle \, d(E_\lambda f, g)$$

where $\hat{G}$ is the dual of G and E is the spectral measure of unitary extension of the semigroup of isometries corresponding to $X(\cdot)$ - such extension exists by Ito theorem [10]. Hence, for part $X'(\cdot)$ of decomposition of Th. 2.0 (2.4) applies. This gives among others an explantation through geometry of the process the appearing in decompositions given in [11] of some parts depending on (t-s).

In all what has been told above we required that involved semigroups of contractions have unitary dilations. That this is the case for S in Cor. 2.1 follows from a generalization of Sz.-Nagy theorem [24], proved in [17].

(2.5) Let $S = \{t \geq 0 : t \in G\}$ where G is an additive subgroup of reals. Then every unital semigroup of contractions over S has a unitary dilation.

Taking $S = \mathbb{Z}^+$ we get therefore the original theorem of Sz.-Nagy. Taking $S = \mathbb{R}_+$ and weakly continuous semigroup of contractions we get next theorem of Sz.-Nagy.

Let $S_+ = \mathbb{Z}^+ \times \mathbb{Z}^+$ be the additive semigroup of latlice points. Then the theorem of Ando [2] states that

(2.6) Every unital semigroup of contractions over S_+ has a unitary dilation.

The "continuous" version of Ando's theorem has been proved by Słociński [23]:

(2.3) Every weakly continuous unital semigroup of contractions over the semigroup $\mathbb{R}_+ \times \mathbb{R}_+$ has a unitary dilation.

Notice that the celebrated example of Parrott [22] shows that a semigroup of contractions over $S = \mathbb{Z}^+ \times \mathbb{Z}^+ \times \mathbb{Z}^+$ need not have a unitary dilation.

The most complete theory of semigroup of contractions is that one of Sz.-Nagy - Foiaş [25], for $S = \mathbb{Z}^+$ and $S = \mathbb{R}_+$ - continuous semigroups. For other references see the expository paper [19].

3. In what follows we assume that (2.0) holds true and the assumptions of Th. 2.0 are satisfied. Let us assume additionally that G is a locally compact topological group and let $\hat{G}$ be its dual. Then every unitary representation $U(\cdot)$ of G can be written as

$$U(t) = \int_{\hat{G}} < t,\lambda > dE_\lambda \ , \ t \in G$$

where E is a spectral measure and $< t,\lambda >$ the value of character λ at $t \in G$. Hence, if $U(\cdot)$ is a unitary dilation of a semigroup of contractions $T(t) \in L(\hat{H})$ over S then for $t \in S$, $f,g \in \hat{H}$

$$(T(t)f,g) = \int_{\hat{G}} < t,\lambda > d(E_\lambda f,g)$$

Suppose $\hat{H} = H_C$ and let $T(\cdot)$ be a unital semigroup of contractions of the process $X(\cdot)$ with covariance $C(\cdot,\cdot)$.

It follows that

$$(3.0) \qquad T(t) = \int_{\hat{G}} < t,u > dF_u, \quad t \in S$$

where F is a semi-spectral measure on Borel subset of $\hat{G}$, namely $F(\sigma)f = PE(\sigma)f$ for $f \in H_c$ and Borel set $\sigma \subset \hat{G}$. In case (2.5) F is determined uniquely by (3.0). In cases (2.6) and (2.7) this is no more true in general. Suppose now that $f \in H_c$. Then $T(t)f = \int_{\hat{G}} < t,u > df_u$ $(t \in S)$ where $f(\sigma) = F(\sigma)f$, the integral being a strong one. If H_c is interpreted as a subspace of $L^2(P)$, where P is a probability space, then $f(\sigma)$ becomes a random measure. Such interpretation is always possible, since H_c can be taken to be a gaussian space. Certainly $H_c \subset L^2(P)$ if originally $X(t)f \in L^2(P)$ i.e. $X(.)$ is just the given process. Then by (3.0)

$$(3.1) \qquad X(t)f = T(t)f = \int_{\hat{G}} < t,\lambda > df_\lambda$$

for $t \in S$, $f \in H$. We then say that $X(.)$ has a semi-spectral representation. (3.0) and (3.1) are analogons of spectral representations of stationary processes. We conclude that every dissipative process $X(.)$ with covariance $C(.,.)$ such that $C(0,0) = I$, over S appearing in (2.5)-(2.7) has a semi-spectral representation. It is obvious that such process is harmonizable in sense of Loeve-Rosanov.

The properties of $T(.)$ imply properties of F appearing in (3.0) and $f(\sigma)$ of (3.1). We refer here to [16] and [25] for basic results in this subject and to [15],[18],[20] for other results. They can be applied directly to processes. Here is an example modeled other a theorem of [18].

<u>Example.</u> Suppose we are given a covariance $c(t,s)$ $(t,s \geq 0)$ which is dissipative and continuous. Suppose that $c(0,0) = 1$. Then by Th. 1.0 there is a strongly continuous unital semigroup $T(t)$, $t \geq 0$ such that

$$(T(t)f_0, T(s)f_0) = c(t,s), \quad t,s \in \mathbb{R}_+ ,$$

where $f_0 = 1$ (dim $H = 1$ in this case). Suppose

$$||T(t)f_0||^2 = c(t,t) \underset{t\to+\infty}{\longrightarrow} 0$$

Then $T(t) \in L(H_c)$ is completely nonunitary. By Sz.-Nagy theorem [24] $T(.)$ has a unitary dilation $U(.)$ say

$$T(t)f = PU(t)f, \quad t \geq 0, \quad f \in H_c \quad \text{where}$$

$$U(t) \in L(K), \quad K \supset H_c$$

$U(t)$ is strongly continuous. We take $f \in H_c$ and define

$$M_\pm(f) = \bigvee_{t \geq 0} U(\pm t)f \quad \text{and}$$

$$R_\pm(f) = \bigcap_{t \geq 0} U(\pm t)M_\pm(f)$$

Notice that $R_+(f) = R_-(f)$.
It follows from Prop.2.1 that $R_+(f) = R_+(f) \cap R_-(f)$ is contained in the space which reduces $T(\cdot)$ to a unitary representation of $\mathbb{R}_+$. Since $T(\cdot)$ is c.n.u. we have

(3.2) $$R_+(f) = R_+(f) \cap R_-(f) = \{0\}$$

Let $\mu = (Ef,f)$ where E is the spectral measure of $U(\cdot)$, that is

$$U(t) = \int_{-\infty}^{+\infty} e^{it\lambda} dE_\lambda \ , \quad t \in \mathbb{R}^1 ,$$

by Stone's theorem. Then the process $f(t) = U(t)f$ is stationary and its spectral domain is $L^2(\mu) = L^2.U(t)|L^2$ is unitarily equivalent to multiplication by $e^{it\lambda}$, f corresponds to $\hat{f}(\lambda) \equiv 1$, and by (3.2) $f(t)$ is completely nondeterministic. By classical theory of stationary processes - see [4] for instance

(3.3) μ is absolutely continuous with respect to linear Lebesque measure m on $(-\infty, +\infty)$.

and

(3.4) The integral $$\int_{-\infty}^{+\infty} \frac{|\log \frac{d\mu}{dm}|}{1+\lambda^2} dm_\lambda$$ is finite.

(3.3) is an analogon of a result of Sz.-Nagy-Foiaş [25], (3.4) an analogon of a results of [15]. We take now $f = f_0$. Then for $\mu = \mu_0 = (Ef_0, f_0)$

$$c(t,0) = (T(t)f_0, f_0) = \int_{-\infty}^{+\infty} e^{it\lambda} d\mu_0(\lambda)$$

for all $t \geq 0$ where μ_0 satisfies (3.3) and (3.4). We thus see that $c(t,0)$ is a Fourier transform of a pretty nice measure.

4. The present section concerns dissipative one dimensional processes with discrete time. This means that we are given a sequence $f_0, f_1, f_2, f_3 \ldots$ of second order random variables with

dissipative covariance. More precisely, if we define

$$C_{n,m} = (f_n, f_m) \qquad n,m = 0,1,2,\dots$$

then for every p and every $\lambda_1 \dots \lambda_p \in \mathbb{C}$

$$\sum_{i,k=0}^{p} C_{i+1,k+1}\lambda_i\bar{\lambda}_k \leq \sum_{i,k=0}^{p} C_{i,k}\lambda_i\bar{\lambda}_k . \tag{4.0}$$

We assume that

$$(f_o, f_o) = \|f_o\|^2 = 1 \tag{4.1}$$

By previous theorems there is a unique contraction T in the space H_c of the proces such that $f_n = T^n f_0$ for $n = 0,1,\dots$.
H_c is spanned by vectors $f_0, f_1 \dots$ which means that T is cyclic with cyclic vector equal to f_0. The covariance can be written as

$$c_{n,m} = (T^n f_0, T^m f_0) \qquad n,m = 0,1,2,\dots$$

Denote by Δ_n the determinant

$$\begin{vmatrix} C_{0,0}, & C_{0,1} & \cdots & C_{0,n-1}, & C_{0,n} \\ C_{1,0}, & C_{1,1} & \cdots\cdots & & C_{1,n} \\ \cdots & \cdots & \cdots & \cdots & \cdots \\ \cdots & \cdots & \cdots & \cdots & \cdots \\ \cdots & \cdots & \cdots & \cdots & \cdots \\ C_{n,0}, & C_{n,1} & \cdots\cdots & & C_{n,n} \end{vmatrix}$$

Since the matrix $(C_{n,m})$ is positive definite, we have $\Delta_n \geq 0$ for all n and $\Delta_0 > 0$ by (4.1).

Suppose now that

$$\Delta_m = 0 \quad \text{for some} \quad m \geq 1 \tag{4.2}$$

and let $n = \min\{m : \Delta_m = 0\}$.

Since Δ_k is the Gramm determinant of $f_0, Tf_0, \dots T^k f_0$, it follows that the vectors $f_0, T f_0, \dots T^{n-1} f_0$ are lineary independent and $f_0, Tf_0, \dots, T^n f_0$ are lineary dependent. It follows that

$$(a_0 I + a_1 T + \dots + a_n T^n) f_0 = 0$$

for some $a_0, \dots, a_n$ such that $\sum_{i|0}^{n} |a_i| > 0$. We must have $a_n \neq 0$, for otherwise $f_0,\dots,T^{n-1}f_0$ would be dependent. Hence

$$-\frac{1}{a_n}(a_0 I + a_1 T\dots a_{n-1}T^{n-1})f_0 = T^n f_0.$$

Denoting by $[h_1,\dots,h_p]$ the linear span of vectors $h_1\dots h_p$ we conclude that

$$T^n f_0 \in [f_0, Tf_0,\dots,T^{n-1}f_0]$$

and consequently, by easy induction

$$T^{n+k}f_0 \in [f_0, Tf_0,\dots,T^{n-1}f_0]$$

for $k = 0,1,2\dots$. It follows that $\dim H_c = n$, and $\{f_0, Tf_0 \dots T^{n-1}f_0\}$ is a linear basis of H_c. The matrix corresponding to T in this basis is defined as $t_{ik} = (Th_k, h_i)$ where $h_j = T^j f_0$ $j = 0,\dots,n-1$. Hence $t_{ik} = (T^{k+1}f_0, T^i f_0) = C_{k+1,i}$ for $k,i = 0,\dots,n-1$ which means that covariance determines directly T. We can now apply the methods of linear algebra to investigate T through matrix (t_{ik}).

We are not interested in doing this and in what follows we focus our attention on the case, when H_c is infinite dimensional. It is obvious that $\dim H_c = +\infty$ if and only if $\Delta_n > 0$ for all $n = 0,1,2,\dots$. We see thus that $\dim H_c$ can be determined via Δ_n by covariance function.

We conclude that

(4.3) $\qquad \dim H_c = +\infty$ iff $\Delta_n > 0$ for $n = 0,1,2\dots$.

If we restrict to processes without conservative part we have to assume that

(4.4) $\qquad (T^n f_0, T^n f_0) = C_{n,n} \underset{n\to\infty}{\longrightarrow} 0$

Since f_0 is cyclic for T, (4.4) says that

(4.5) $\qquad T^n h \underset{n\to\infty}{\longrightarrow} 0 \quad$ for $h \in H_c$

In the study of our T we follow partly the Sz.-Nagy-Foiaş theory [25]. The basic role is played here by the operators

$$D_T = (I - T^*T)^{\frac{1}{2}}, \quad D_{T^*} = (I - TT^*)^{\frac{1}{2}}$$

The simplest cases are when closures of ranges of these operators have finite dimensions. This cases are characterized by a refinement of Th. 6.2 [13] in a lemma below.

We do not assume that (4.4) holds true. If (4.4) holds true then $\dim \overline{R(D_{T*})} \leq \dim \overline{R(D_T)}$

<u>Lemma 4.0.</u> Let A be a bounded positive operator in the Hilbert space K and suppose that the sequence $h_1, h_2, h_3 \ldots$ spans K i.e. linear manifold spanned by these vectors is dense in K. Then the closure $\overline{R(A)}$ of the range $R(A)$ is finite dimensional if and only if there is a finite integer r such that for every n, every $p_1 \ldots p_n$ and $h_{p_1} \ldots h_{p_n}$ the rank $r(h_{p_1} \ldots h_{p_n})$ of the matrix $\{(Ah_{p_i}, Ah_{p_k})\}_{i,k=1\ldots n}$ satisfies the inequality $r(h_{p_1} \ldots h_{p_n}) \leq r$. If this is the case then

$$(4.6) \qquad r_A \overset{df}{=} \dim \overline{R(A)} = \max_n r(h_{p_1} \ldots h_{p_n}) \quad (1 \leq p_i < +\infty)$$

<u>Proof.</u> If $r_1 < +\infty$ then for $n = r_A + 1$ the vectors $Ah_{p_1} \ldots Ah_{p_n}$ are lineary dependent which implies that their Gramm determinant is zero. Since $p_1 \ldots p_n$ are arbitrary then $r(h_{q_1} \ldots h_{q_m}) \leq r_A$ for $m > r_A$. Suppose now that $r(h_{p_1} \ldots h_{p_n}) \leq r$ for some r, arbitrary n and arbitrary positive integers p_i. Hence

$$r(h_1 \ldots h_n) \leq r$$

for every n. It follows that

$$\dim [Ah_1, \ldots, Ah_n] \leq r$$

This is obvious if $n \leq r$. So suppose $n > r$ and $\dim [Ah_1 \ldots Ah_n] > r$. Let $e_1 \ldots e_q$ be the orthonormal basis of $[Ah_1 \ldots Ah_n]$. Then

$$\operatorname{rank} ((Ah_i, e_k))_{\substack{i=1\ldots n \\ k=1\ldots q}} = \dim[Ah_1 \ldots Ah_n] = d$$

Changing the order of h_i and e_k if necessary we can assume that

$$Ah_i = \sum_{j=1}^{d} x_{ij} e_j, \qquad i = 1, \ldots, d$$

and the determinant $\det(x_{ij})_{ij=1\ldots d} \neq 0$.

Let $a_{ik} = (Ah_i, Ah_k)$. Then

$$a_{ik} = \sum_{j=1}^{d} x_{ij} \overline{x_{kj}} \quad \text{for } i = 1 \ldots d,$$

which means that the matrix $(a_{ik})_{i,k=1\ldots d}$ is the Cauchy product of (x_{ij}) and $(\overline{x_{kj}})$. Hence

$$\det (a_{ik}) = \det (x_{ij}) \det (\overline{x_{kj}}) = \det (x_{ij}) \det (\overline{x_{jk}}) =$$
$$\det (x_{ij}) \overline{\det (x_{jk})} = |\det (x_{pq})|^2 > 0.$$

It follows that $r(h_1 \ldots h_n) = d > r$ which is a contradiction. We proved in this case that $\dim \overline{R(A)} \leq r$, which by the first part of the proof completes the proof of right hand equality of (4.6). Using our Lemma 4.0. one can formulate a necessary and sufficient conditions for finite dimensionality of the closure of the range of D_T where T is a contraction related to dissipative process over $\mathbb{Z}^+$ with initial space H not necessarily one or even finite dimensional. If $\dim H = 1$ as in the case we actually consider, we derive from Lem. 4.0 the following.

Proposition 4.0. Let $\{c_{n,m}\}$ be a dissipative matrix and T the related contraction (i.e. $c_{n,m} = (f_n, f_m)$ where $T^i f_0 = f_i$, $\|f_0\| = 1$). Denote $d_{i,k} = c_{i,k} - c_{i+1,k+1} = (D_T T^i f_0, D_T T^k f_0)$. Then $\delta_T = \dim \overline{D_T H_c} = 1$ if and only if every minor of second order the matrix $(d_{ik})_{i,k=1,2\ldots}$ equals to zero.

The above proposition expresses in terms of covariance quite effectively the property that $\delta_T = 1$. We assume since now that

(4.7) $\qquad \delta_T = 1$

If (4.4) i.e. (4.5) holds also true, then by Prop. 2.1, I[25] $\delta_{T*} = \dim \overline{D_{T*} H_c} \leq 1$. Hence two cases are posible :

(4.8) $\qquad \delta_{T*} = 0$

or

(4.9) $\qquad \delta_{T*} = 1$

If (4.8) folds true, then T^* is an isometry. Since $\delta_T = 1$ we must have $\dim H_c = +\infty$. Denote $S = T^*$. Then $\dim \overline{(I - SS^*)H_c} = \delta_T = 1$. On the other hand SS^* is an orthogonal projection on SH_c. It follows that $(I - SS^*)$ is the orthogonal projection on the wandering subspace $\Delta = H_c \ominus SH_c$, $\dim \Delta = 1$ and, since $S^{*n} \longrightarrow 0$ strongly we have $H_c = \bigoplus_{n|0}^{\infty} S^n \Delta$ provided (4.4) holds true. This means that S is a uniteral shift of multiplicity one. Hence T is a backward shift. The map $S^n e \longrightarrow z^n$

(e - a unit vector whose multiples are all of Δ) establishes the unitary equivalence between H_c and the Hardy space H^2. There corresponds to $S^* = T$ the backward shift $\hat{S}^*$ which maps $h(\cdot) \in H^2$ onto function $\frac{h(z)-h(0)}{z} = (S^*h)(z)$. Next, since $(S^n S^{*n} - S^{n+1} S^{*n+1})$ is the orthogonal projection onto $S^n \Delta$ then

$$f_o = \sum_{n|0}^{\infty} (f, S^n e) S^n e$$

Suppose for simplicity that $f_o - SS^* f_o \neq 0$ that is $||Tf_o||^2 = c_{11} < c_{00} = ||f_o||^2 = 1$.

Then we can take

$$e = \frac{f_o - SS^* f_o}{||f_o - SS^* f_o||}$$

and then get for $n = 0,1,2,\ldots$

$$(f, S^n e) = \frac{(T^n f_o, f_o) - (T^{n+1} f_o, Tf_o)}{\sqrt{1 - ||Tf_o||^2}} = \frac{c_{n,o} - c_{n+1,1}}{\sqrt{1 - c_{11}}}$$

It follows that the Taylor coefficients of the function $\hat{f}_o \in H^2$ which corresponds to f_o can be determined by covariance. $\hat{f}_o(\cdot)$ is a cyclic vector for $\hat{S}^*$. Conversely, if we are given a unit cyclic vector $\hat{f}_o \in H^2$ for $\hat{S}^*$, then we can define a dissipative matrix $\{\hat{c}_{n,m}\}$ with entries

$$\hat{c}_{n,m} = (\hat{S}^{*n} \hat{f}_o, \hat{S}^{*m} \hat{f}_o).$$

The corresponding contration T will be then $\hat{S}^*$ and obviously satisfies (4.5), (4.7) and (4.8) . It follows that there is a one-to-one correspondence between processes $\{f_n\}$ satisfying these three conditions and processes created with the help of $\hat{S}^*$ and its cyclic vectors, which correspond to f_o . Since, for instance the function

$$\hat{f}_o(z) = \sum_{k|0}^{\infty} \frac{1}{k+1} z^k$$

is a cyclic vector for $\hat{S}^*$, see [5] , then the matrix with entries

$$c_{n,m} = \rho \sum_{k|0}^{\infty} \frac{1}{k+n+1} \cdot \frac{1}{k+m+1} \quad ; \quad n,m = 0,1,\ldots$$

where ρ is a normalizing factor such that $c_{oo} = 1$, will be a covariance such that (4.5) holds true and the corresponding T

satisfies (4.7) and (4.8). In this way we get a concrete example of dissipative covariance in case when $\delta_{T*} = 0$, $\delta_T = 1$ and when (4.5) holds true.

Next we will consider the case when (4.5), (4.7) and (4.9) hold true. Then using Th. 4.1. I and Th. 2.3. VI of [25] or a direct approach of de Branges - Rovnyak [6] Th. 12, and Beurling invariance theorem *) or the approach of Helson [9] we derive then that for some inner function $\varphi \in H^2$, H_c can be identified with $H^2 \ominus \varphi H^2$ and T with restriction of the backward shift S^* in H^2 to $H^2 \ominus \varphi H^2$ i.e. $T^n = S^{*n}|H^2 \ominus \varphi H^2 = H_\varphi$, $n \geq 0$.

To point out the depedence of T on φ we write T_φ in place of T. The concrete models of T_φ are described in [1]; they are completed in detail by Nikolski in [21]. Those models express effectively T_φ thorough parameters of φ induced by Nevanlinna factorization. Let namely $\varphi = B\, s\, \Delta$ be this factorization where B is a Blaschke product, s a singular function with continuous measure and Δ with an atomic one. Then - see [1], [21]

(4.10) $$H_\varphi = H_B \oplus BH_s \oplus B\, s\, H_\Delta$$

and T_φ can be expressed as an operator triangular matrix with to decomposition (4.10). We are interested in determination of covariance $\{c_{n,m}\}$ in terms of φ and the presise form of T_φ will be not needed. For illustration we restrict ourselves to the case when $\varphi = B$. The crucial points here, presented in [1], is the determination of the basis of $H_c = H_B$ in terms of factors of B. In what follows we apply the methods of [1].

Let $|a_1| \leq |a_2| \leq |a_3| \leq \dots$ be all zeros of B with multiplicietis included. Then

$$B(z) = \prod_{n|1}^{\infty} \left(- \frac{\bar{a}_n}{|a_n|} \right) \frac{z-a_n}{1-\bar{a}_n z}$$

where $\frac{\bar{a}_n}{|a_n|}$ is understood as 1 if $a_n = 0$. Then, defining

$K_a(z) = \frac{1}{1-\bar{a}z}$ we get that the sequence

*) We refer here to [8], [9] for details.

$$h_1(z) = (1 - |a_1|^2)^{\frac{1}{2}} K_{a_1}(z)$$

$$h_2(z) = (1 - |a_2|^2)^{\frac{1}{2}} \prod_{m|1}^{1} \frac{z-a_m}{1-\bar{a}_m z} K_{a_2}(z)$$

....

....

....

$$h_n(z) = (1 - |a_n|^2) \prod_{m|1}^{n-1} \frac{z-a_m}{1-\bar{a}_m z} K_{a_n}(z)$$

....

....

....

is an onthonormal basis of H_B. In particular, if B has a finite number of factors, then H_B is finite dimensional and we then have to do with the situation discussed at the begining of the present section. We have to exclude the case $B \equiv$ const, because $\delta_{T^*_B} = 1$ in our case.

Let us write now $T_B = T$ for convenience.

We will find out the explicit form of the operator $(I - T^*T) = D_T^2$.

Let us take namely $f(.) \in H_B = H^2 \ominus BH^2$ and its Taylor expansion

$$f(z) = \sum_{n|0}^{\infty} a_n z^n$$

Since $T f = S^*f$ then

$$(Tf)(z) = \sum_{n|0}^{\infty} a_{n+1} z^n$$

Define now $g = P_B u_0$ where P_B is the orthogonal projection of H^2 onto H_B and $u_0(z) \equiv 1$. Suppose $h \in H_B$, $h(z) = \sum_{n|0}^{\infty} b_n z^n$. Then $(T^*Tf,h) = (SS^*f,h) = (S^*f,S^*h) = \sum_{n|0}^{\infty} a_{n+1}\bar{b}_{n+1} =$

$= \sum_{n|0}^{\infty} a_n\bar{b}_n - a_0\bar{b}_0 = (f,h) - a_0\bar{b}_0$. But $b_0 = (h,u_0) = (P_Bh,u_0) = (h,P_Bu_0) = (h,g)$. It follows that $(T^*Tf,h) = (f,h) - a_0\overline{(h,g)} = (f - a_0g,h)$ and since h is arbitrary, then

$$T^*Tf = f - a_0g$$

that is $(I - T^*T)f = a_0 g = (f,u_0)g = (P_B f,u_0)g = (f,P_B u_0)g = (f,g)g$ and consequently

(4.11) $$D_T f = \frac{1}{\mu}(f,g)g \quad \text{for} \quad f \in H_B$$

where $\mu = ||g||$.

Notice now, that since $T = S^*|H_B$ then $T^{*n}f = P_B S^n f$ for $n = 0,1,\dots, f \in H_B$. Using the special form of the basis $\{h_n\}$ one proves easily that

(4.12) $$(T^{*k}h_n,h_m) = \int_\Gamma z^k h_n \bar{h}_m d\mu = (h_n,T^k h_m) = 0$$

if $n > m \geqslant 1$, for $k = 0,1,2,\dots$ ($\Gamma = \{z | |z| = 1\}$, μ = the normalized Lebesque measure on Γ). This is the triangularity property of T with respect to the basis $\{h_n\}$. We define now for $f \in H_B$

$$\Delta(m,n;f) = (D_T T^m f, D_T T^n f)$$

Then, by (4.11)

$$\Delta(m,n;f) = (T^m f,g)\,\overline{(T^n f,g)}$$

Let $L^2_B = \mathbb{C}^p$ is $\dim H_c = p < +\infty$ and $L^2_B = l^2$ if $\dim H_c = +\infty$.

The map $f \longrightarrow \{a_n\} \in L^2_B$ where $f(z) = \sum_{n\geqslant 1} a_n h_n(z)$, is a unitary one of H_B onto L^2_B. We have

$$T^k f = \sum_{m\geqslant 1} a_m T^k h_m$$

On the other hand

$$g = \sum_{n\geqslant 1} g_n h_n$$

Hence

$$(T^k f,g) = \sum_{m\geqslant 1} a_m (T^k h_m,g) = \sum_{m\geqslant 1} a_m \Big(\sum_{n\geqslant 1} \bar{g}_n (T^k h_m,h_n)\Big)$$

which by (4.12) proves that

$$(T^k f,g) = \sum_{m\geqslant 1} a_m \sum_{n|1}^{m} (T^k h_m,h_m) .$$

We define now auxiliary functions

$$u_m(k) = \sum_{n|1}^{m} \bar{g}_n \quad (T^k h_m,h_n)$$

Then

$$(T^k f,g) = \sum_{m\geq 1} a_m u_m(k)$$

Since $g_n = (g,h_n) = (P_B u_0, h_n) = \int_\Gamma \overline{h}_n \, d\mu$ the coefficients g_n are determined uniquely by $a_1 \dots a_n$ and can be computed explicitely. Next, since

$$(T^k h_m, h_n) = \int_\Gamma h_m \overline{z}^k \overline{h}_n d\mu \quad (m \geq n)$$

then $n \leq m$ in formula

$$u_m(k) = \sum_{n|1}^{m} \overline{g}_n \int_\Gamma h_m \overline{z^k h_n} d\mu \ ;$$

hence $u_m(\cdot)$ is determined uniquely by $a_1 \dots a_n$ merely, and can be expressed via $a_1 \dots a_m$ explicitely. Now

$$\Delta(k,j) = \Phi_f(k) \ \overline{\Phi_f(j)}$$

where

$$\Phi_f(i) = (T^i f, g) = \sum_{m\geq 1} a_m u_m(i)$$

On the other hand, since $T^n f \xrightarrow[n\to\infty]{} 0$ we have

$$(T^m f; T^n f) = \sum_{k|0}^{\infty} [(T^{m+k} f, T^{n+k} f) - (T^{m+k+1} f, T^{n+k+1} f)] =$$

$$\sum_{k|0}^{\infty} (D_T T^{m+k} f, D_T T^{n+k} f) = \sum_{k|0}^{\infty} \Phi_f(m+k) \ \overline{\Phi_f(n+k)}$$

Summing up, when knowing the coordinates of $f \in L_B^2$, identifying L_B^2 with H_B as above, we are able to compute thorough them and by $a_1, a_2 \dots$ the number $(T^m f, T^n f)$. This together with the above formulae is the full analogon of Th. 7.1, 7.2 of [13].
Let us point out that if the number 1 is in the closure of zeros $\{a_n\}$, then T will be a cogenerator of a one parameter continuous semigroup of contractions over $\mathbb{R}_+$ with unbounded generator.
So, apriori we can investigate processes with continuous time, which correspond to semigroups with unbounded generators - see [16], [25] for details.

A few final general comments are now in order. First of all notice that we left behind us the question which properties of covariance $\{c_{n,m}\}$ decide whether (4.8) or (4.9) holds true. The problem arises to find out reasonable conditions in terms of covariance matrix. We described in some detail the case $\delta_T = \delta_{T*}=1$. Also here the question appears, namely how to determine φ by covariance - and a more difficult problem arises, namely when φ reduces to one of canonical factors: for example, when φ is a Blaschke product. Similar problems arise if δ_T, δ_{T*} are finite, but greater than one. Then φ is replaced by the characteristic function of T and a problem appears how to find out the functional model in terms of covariance or how to determine covariance through parametres which form a complete set of unitary invariants of T.

References

[1] Ahern P.R. and Clark D.N. - On functions orthogonal to invariant subspaces, Acta Math. 124 (1970) 191-204.

[2] Ando T. - On a pair of commutative contractions, Acta Sci. Math. 24 (1963) 88-90.

[3] Cramer H. - On some classes of nonstationary stochastic processes. Proc. of IV Berkeley Symp. on Statistic and Appl. Probability II, (1960) 57-78.

[4] Cramer H. and Leadbetter M.R. - Stationary and stochastic processes, New York, 1969.

[5] Douglas R.G., Shapiro H.S. and Shields A.L. - Cyclic vectors and invariant subspaces for the back ward shift operator, Ann. Inst.Fourier Tome XX, Fasc. 1 (1970) 37-76.

[6] de Branges L. and Rovnyak J. - Square summable power series, New York 1966.

[7] Getoor R.K. - The shift operator for nonstationary stochastic processes, Duke Math. J. Vol. 23 (1956) 175-187.

[8] Hoffman K. - Banach spaces of analytic functions, Englewood Cliffs N.J. 1962.

[9] Helson H. - Lectures on invariant subspaces, New York - London, 1964.

[10] Ito T. - On the commutative family of subnormal operators, J. Fac.Sci. Hokkaido Univ. (1) 14 (1958) 1-15.

[11] Kircev K.P. - On a certain class of non-stationary random processes. Teoria funkci, funkc.analis (Harkov 1971) vyp.14 150-169 (in Russian).

[12] Livsic M.S. and Brodski M.S. - Spectral analysis of non-seladjoint operators and intermediate systems, Amer. Soc. Transl. Ser. 2. 13 (1960) 265-346.

[13] Livsic M.S. and Jancevic - Theory of operator nodes in Hilbert spaces, Harkov 1971 (in Russian)

[14] Loève M. - Probability theory, New York 1955.

[15] Mlak W. - Characterization of completely non-unitary contractions in Hilbert spaces, Bull. Ac.Sc. Polon. 12 (1963) 111-114.

[16] ———, - Unitary dilations of contraction operators, Rozprawy Matem. 46 (1965) 1-88, Warszawa.

[17] ———, - Unitary dilations in case of ordered groups, Ann. Pol. Math. 17 (1965) 321-328.

[18] ———, - On semi-groups of contractions in Hilbert spaces, Studia Math. 26 (1966) 263-272.

[19] ———, - Operator valued representations of function algebras, Linear operators and Approx. II. 1974, 49-79.

[20] Muhly P.S. - Some remarks on the spectra of unitary dilations, Studia Math. XLIX (1974) 139-147.

[21] Nikolski N.K. - Lectures on shift operator (Lecture 8) Zapiski naucnik seminarov LOMI, Vol. 56, Leningrad 1976, 104-127.

[22] Parrott S. - Unitary dilations for commuting contractions, Pacific Math. J. Vol. 34. 2. (1970) 481-490.

[23] Słociński M. - Unitary dilations of two parameter semigroup of contractions, Bull. Ac.Sc. Polon. 22, 10 (1974) 1011-1014.

[24] Sz.-Nagy B. - Sur les contractions de l'espace de Hilbert, Acta Sci. Math. 15 (1953) 87-92.

[25] Sz.-Nagy B. and Foiaş C. - Harmonic analysis of operators on Hilbert space. Akadem. Kiado, Budapest 1970.

[26] Weron A. - Prediction theory in Banach spaces. Prob. Winter School.(Karpacz 1975). Springer Lect.Notes in Math. 472 (1975), 207-228.

[27] ——, - On weak second order and gaussian random elements. Probability in Banach spaces, (Oberwolfach 1975. Springer Lect.Notes in Math. 526 (1976) 263-272.

Instytut Matematyczny PAN
Oddział w Krakowie
31-027 K r a k ó w
ul. Solskiego 30
P o l a n d

EXAMPLES OF NON-STATIONARY BANACH SPACE VALUED STOCHASTIC PROCESSES OF SECOND ORDER

Nguyen Van Thu and A. Weron

In [8] Banach space valued stochastic processes of second order as curves in a Loynes space were studied. Here we give further two examples of non-stationary second order processes related to additive correlated random variables and Markov processes. The aim of this paper is to present a general form of processes of these two classes defined over full LCA semigroups and semilattices, respectively. Moreover we obtain representations of operator valued positive-definite functions on the above mentioned semigroups.

0. If B_1, B_2 are Banach spaces then $L(B_1, B_2)$ denotes the space of all continuous linear operators from B_1 into B_2. Let B be a Banach space and H a Hilbert space. It is known [8] that the space L(B,H) is a Loynes space with the operator valued inner product

$$[X,Y] = Y^*X \in L(B,B^*),$$

where B^* denotes the topological dual of the space B. By a Banach space valued (B-valued) process we mean a family $(X_t)_{t\in T}$ of elements of L(B,H) indexed by a set T, (cf. [1] and [8]). Its correlation function is defined as $K(t,s) = [X_t, X_s]$. Let H_X denotes the vector-time-domain of the B-valued process $(X_t)_{t\in T}$

$$H_X = \text{span}\{X_t b,\ b \in B,\ t \in T\} \subset H$$

Let S be an Abelian semigroup with the multiplication "$\circ$". A B-valued process $(X_t)_{t\in S}$ is said to be <u>m-correlated</u> if for any $t,s \in S$ $K(t,s) = K(t \circ s)$.

In particular if $(\xi_t)_{t\in S}$ is a real valued stochastic process such that $E\,\xi_t^2 < \infty$, $t \in S$, then it is called m-correlated if for any $t,s \in S$

$$E\,\xi_t\,\xi_s = E\,\xi_{t\,\circ\,s}\,\xi_e$$

Such processes were studied in [5] as additively correlated sequences of random variables over the additive semigroup Z_+ of non-negative integers. In the earlier paper [2] Getoor considered such stochastic processes for the case of the additive semigroup R_+ of non-negative reals and called it "symmetric random functions". But his starting point was the shift operator.

Lemma 0.1. Let $(X_t)_{t\,\varepsilon\,S}$ be a B-valued m-correlated process. If $X_{t_1}b_1 = X_{t_2}b_2$ for some $t_1,t_2\ \varepsilon\ T$ and $b_1,b_2\ \varepsilon\ B$ then for each $s\ \varepsilon\ S$

$$X_{s\,\circ\,t_1}b_1 = X_{s\,\circ\,t_2}b_2\ .$$

Proof. Let $s,u\ \varepsilon\ S$ and $d\ \varepsilon\ B$. Then we have

$$(X_u d, X_{s\,\circ\,t}b_1) = (X_{s\,\circ\,u}d, X_{t_1}b_1) = (X_{s\,\circ\,u}d, X_{t_2}b_2) = (X_u d, X_{s\,\circ\,t_2}b_2)$$

Since the set $\{X_u d,\ u\,\varepsilon\,S,\ d\,\varepsilon\,B\}$ is linearly dense in H_X we get $X_{s\,\circ\,t_1}b_1 = X_{s\,\circ\,t_2}b_2$.

Given $s\,\varepsilon\,S$ define a linear operator T_s on H_X by the formula

$$T_s(X_t b) = X_{t\,\circ\,s}b \qquad t\ \varepsilon\ S,\quad b\ \varepsilon\ B.$$

By Lemma 0.1. T_s is well defined. Moreover

(i) T_s is self-adjoint,

(ii) $T_sT_t = T_{t\,\circ\,s}$

(iii) T_s is defined on a linearly dense subset of H_X

1. An Abelian topological semigroup S is said to be <u>locally compact full</u> (LCF), compare [7], if

a) S can be embedded in a locally compact group G,

b) S is locally compact in the relative topology of G,

c) every non-empty bounded open set in S has non-zero measure with respect to the Haar measure on G,

d) there exists a denumerable set $D = \{x_n\}$ in S such that if x is any element of S then there exists an element y in S and an element x_n in D such that $x_n = xy$. That is for every $x \in S$ $(xS) \cap D \neq \emptyset$.

Let $(X_t)_{t \in S}$ be a B-valued m-correlated process over a LCF Abelian semigroup S. By virtue of Theorem [6] in [7] the semigroup (T_s) of operators on H_X can be represented in the form

$$T_s = \int_{\Gamma} < s,\gamma > E(d\gamma) ,$$

where Γ is the set of real characters $< \cdot,\gamma >$ of the semigroup S, $E(\cdot)$ is a resolution of identity in H_X. Thus for each $b \in B$

$$X_s b = T_s X_e b = \int_{\Gamma} < s,\gamma > E(d\gamma)\ X_e b .$$

If we put for each Borel set Δ

$$\Phi(\Delta) = E(\Delta) X_e$$

then Φ is an $L(B,H_X)$-valued strongly σ-additive measure on Borel subsets of Γ . Thus we have proved

<u>Theorem 1.1</u>. <u>Every continuous B-valued m-correlated process</u> $(X_t)_{t \in S}$ <u>over a LCF Abelian semigroup S has the following representation</u>

$$X_s = \int_{\Gamma} < s,\gamma > \Phi(d\gamma) , \qquad s \in S,$$

<u>where</u> Φ <u>is an</u> $L(B,H_X)$ <u>valued strongly</u> σ-<u>additive measure on</u> Γ .

<u>Example 1.1</u>. If $S = Z_+^d$, $d = 1,2,\ldots$ then

$$X_n = \int_{R^d} t_1^{n_1}\ t_2^{n_2} \ldots\ t_d^{n_d}\ \Phi\ (dt) ,$$

where $t = (t_1,t_2,\ldots,t_d) \in R^d$ and $n = (n_1,n_2,\ldots,n_d) \in Z_+^d$.

<u>Example 1.2</u>. Let K be a Hilbert space. We say that K-valued process $(Z_n)_{n \in Z_+}$ is <u>biharmonic</u> if

$$[Z_n,Z_k] = \frac{1}{n+k+1}\, I_K \qquad n,k = 0,1,2,\ldots$$

where I_K is the identity operator in K. It is evident that Z_n is m-correlated. Moreover if a sequence of operators $A_n \in L(B,K)$

satisfies the condition that for each $n = 0,1,2,\ldots,$ and $b \in B$ the series

$$\sum_{i,j=0}^{\infty} \frac{(A_i^* A_j b)(b)}{i+j+2n+1}$$

is convergent, then the sequence of "moving average" of the biharmonic sequence Z_n defined by the formula

$$X_n = \sum_{k=0}^{\infty} Z_k A_{k+n} \qquad n = 0,1,2,\ldots$$

is a B-valued m-correlated process over Z_+ .

Example 1.3. Let $(\varphi_t)_{t \in R}$ be a scalar valued stationary process with continuous time such that there exist derivatives (in the sense of square mean) of all orders. Then

$$\xi_n = i^n \left. \frac{d^n \varphi_t}{dt^n} \right|_{t=1} \qquad n = 0,1,2,\ldots$$

is m-correlated sequence over Z_+ .

Theorem 1.2. For each weakly continuous $L(B,B^*)$ valued function $K(s)$ on a LCF Abelian semigroup S the following conditions are equivalent:

(i) K is a positive-definite function on the semigroup S i.e.,

$$\sum_{i,j=1}^{N} (K(t_i \circ t_j)b_i)(b_j) \geq 0$$

for all $N = 1,2,\ldots$, $t_i \in S$ and $B_i \in B$ $i = 1,2,\ldots,N,$

(ii) K is the correlation function of a continuous B-valued m-correlated process over S,

(iii) there exists a positive $L(B,B^*)$-valued weakly σ-additive measure F on Borel subsets of the dual semigroup Γ of S such that

$$K(s) = \int_{\Gamma} < s,\gamma > F(d\gamma) \quad , \qquad s \in S .$$

Proof. (i) $\Rightarrow$ (ii). If K is positive definite in the sense of (i) then $R(t,s) = K(t \circ s)$ is positive definite $L(B,B^*)$-valued function on $S \times S$. Therefore by theorem 4.6 in [8] $R(t,s)$ is the correlation function of a continuous B-valued process of second order $(X_t)_{t \in S}$. But in view of the condition $R(t,s) = K(t \circ s)$ the process $(X_t)_{t \in S}$ is m-correlated.

(ii) $\Rightarrow$ (iii). Proceeding succesively by Theorem 1.1. we have

$$K(s) = K(s \circ e) = [X_s, X_e] = \int_\Gamma < s,\gamma >[\Phi(d\gamma),\Phi(d\gamma)].$$

Putting $F(d\gamma) = [\Phi(d\gamma),\Phi(d\gamma)]$ and taking into account the last equation we get the condition (iii).

(iii) $\Rightarrow$ (i). This implication is obvious.

As a corollary we obtain the following generalization of Hamburger's theorem [3] to the case of operator valued functions.

<u>Corollary 1.1.</u> <u>A sequence of operators</u> $A_n \in L(B,B^*)$ $n=0,1,2,\ldots$ <u>is positive definite in the sense of the condition</u> (i) <u>on the additive semigroup</u> Z_+ <u>if and only if there is a positive</u> $L(B,B^*)$-<u>-valued weakly</u> σ-<u>additive measure</u> F <u>on Borel subsets of the real line such that</u>

$$A_n = \int_{-\infty}^{+\infty} t^n F(dt) \qquad n = 0,1,2,\ldots .$$

2. A partially ordered set $(T,\leqslant)$ is called a <u>semilattice</u> if for any $t,s \in T$ there exists the minimum $t \wedge s \in T$. The semilattice is an Abelian semigroup with "multiplication" $t \circ s = t \wedge s$. Thus a B-valued process $(X_t)_{t \in T}$ over the semilattice $(T,\leqslant)$ is m-correlated if $K(t,s) = K(t \wedge s)$.

<u>Example 2.1.</u> It is known [6] that a real valued Gaussian process $(\xi_t)_{t \in T}$ with the mean zero is Markov i.e.,

$$E\{\xi_t | \xi_u : u \leqslant s\} = E\{\xi_t | \xi_{t \wedge s}\}, \qquad t,s,u \in T$$

if and only if there exist scalar valued functions f and h on T such that for all $t,s \in T$

$$K(t,s) = E\,\xi_t\,\xi_s = f(t \wedge s)h(t)h(s).$$

Consequently, if $K(t,s) \neq 0$ for $t,s \in T$ then process $(\xi_t)_{t\in T}$ is Markov if and only if the process $(\xi_t/h(t))_{t\in T}$ is m-correlated over the semilattice T.

For any $t \in T$ let J_t denote the cone up to t, i.e., $J_t = \{s \in T : s \leq t\}$. Let $\mathfrak{R}$ be the ring generated by all such cones.

Theorem 2.1. Let $K : T \to L(B,B^*)$, then the following conditions are equivalent:

(i) K is positive-definite on the semilattice T i.e.,

$$\sum_{i,j=1}^{N} (K(t_i \wedge t_j)b_i)(b_j) \geq 0$$

for all $N = 1,2,\ldots$, $b_i \in B$ and $t_i \in T$, $i = 1,2,\ldots,N$,

(ii) K is the correlation function of B-valued m-correlated process over the semilattice T.

(iii) there exists an additive positive $L(B,B^*)$ valued measure F on the ring $\mathfrak{R}$ such that for each $t \in T$

$$F(J_t) = K(t).$$

Proof. (i) $\Rightarrow$ (ii). If K is positive-definite function in the sense of (i) then $R(t,s) = K(t \wedge s)$ is positive-definite $L(B,B^*)$--valued function on $T \times T$. Therefore by Theorem 4.6 in [8] $R(t,s)$ is the correlation function of a B-valued process over T. But in view of the condition $R(t,s) = K(t \wedge s)$, the process $(X_t)_{t\in T}$ is m-correlated.

(ii) $\Rightarrow$ (iii). Since the class of all cones J_t is multiplicative there exists a unique additive $L(B,B^*)$ valued set function F defined on $\mathfrak{R}$ such that

$$F(J_t) = K(t), \qquad t \in T.$$

We prove that for each $\Delta \in \mathfrak{R}$ $F(\Delta)$ is positive. For every set $\Delta \in \mathfrak{R}$ the indicator 1_Δ can be represented as

$$1_\Delta = (\sum_{j=1}^{n} e_j 1_{J_{t_j}})^2$$

for some $t_j \in T$ and $e_j = \pm 1$, $j = 1,2,\dots,n$.
Hence for each $b \in B$

$$(F(\Delta)b)(b) = \int 1_\Delta((dF)b)(b) = \int \sum_{i,j} e_i e_j 1_{J_{t_i}} 1_{J_{t_j}}((dF)b)(b) =$$

$$= \sum_{i,j=1}^{n} e_i e_j (F(J_{t_i} \cap J_{t_j})b)(b) = \sum_{i,j=1}^{n} e_i e_j (K(t_i \wedge t_j)b)(b) =$$

$$= \sum_{i,j=1}^{n} (K(t_i \wedge t_j)e_i b)(e_j b).$$

Since the correlation function K is positive-definite thus the last term is positive. Consequently F is a positive L(B,B*) valued additive measure on $\mathfrak{R}$ which satisfies the condition (iii).

(iii) $\Rightarrow$ (i). Let F be as in (iii). If N is natural number, $b_1,\dots,B_N \in B$ and $A_1,\dots,A_N \in \mathfrak{R}$ thus we must show that

$$\sum_{i,j=1}^{N} (F(A_i \cap A_j)b_i)(b_j) \geq 0.$$

Now to the sets $A_1,\dots,A_N$ there correspond disjoint sets $D_1,\dots,D_M$ in $\mathfrak{R}$ such that we have the partioning

$$A_i = \bigcup_{n \in N_i} D_n ,$$

where $\{1,\dots,M\} = \sum_{i=1}^{N} N_i$. It follows that

$$A_i \cap A_j = \bigcup_{n \in N_i \cap N_j} D_n ,$$

where the sets D_n are disjoint. Therefore from the finite-additivity of $F(\cdot)$ we have

$$F(A_i \cap A_j) = \sum_{n \in N_i \cap N_j} F(D_n) = \sum_{k=1}^{M} F(D_k)\, 1_{N_i \cap N_j}(k).$$

Thus

$$\sum_{i,j=1}^{N} (F(A_i \cap A_j)b_i)(b_j) = \sum_{i,j=1}^{N} \sum_{k=1}^{M} (F(D_k)b_i)(b_j)\, 1_{N_i \cap N_j}(k) =$$

$$= \sum_{k=1}^{M} \Big[\sum_{i,j=1}^{N} (F(D_k)b_i\, 1_{N_i}(k))(b_j\, 1_{N_j}(k))\Big] =$$

$$= \sum_{k=1}^{N} (F(D_k)p_k)(p_k) \geq 0 \ ,$$

where

$$p_k = \sum_{i=1}^{N} b_i\, 1_{N_i}(k) \in B,$$

which completes the proof.

Let $(X_t)_{t \in T}$ be a B-valued m-correlated process over the semilattice T. According to Theorem 2.1 there exists an $L(B,B^*)$ valued positive additive measure F on the ring $\mathfrak{R}$ such that

$$\sum_{i,j=1}^{N} (F(J_{t_i} \cap J_{t_j})b_i)(b_j) \geq 0$$

for all natural numbers N, $J_1,\dots,J_N \in \mathfrak{R}$ and $b_1,\dots,b_N \in B$. Let $\mathfrak{X}$ denote the linear manifold of all B valued simple functions f on T of the form

$$f(t) = \sum_{i=1}^{N} b_i\, 1_{D_i}(t), \qquad b_i \in B, \ \ D_i \in \mathfrak{R} \ .$$

Now we define an inner product on $\mathfrak{X}$ as follows

$$(f^1,f^2) = \sum_{i,j=1} (F(D_i^1 \cap D_j^2)b_i^1)(b_j^2) \ .$$

Let $\overline{\mathfrak{X}}$ denote the completion of $\mathfrak{X}$ in this inner product.

<u>Lemma 2.1.</u> $\overline{\mathfrak{X}}$ <u>is isometrically isomorphic with</u> H_X.

Proof. Let $t_1, t_2, \dots, t_N \in T$ and $b_1, b_2, \dots, b_N \in B$. Thus

$$\| \sum_{i=1}^{N} X_{t_i} b_i \|^2_{H_X} = \sum_{i,j=1}^{N} (K(t_i \wedge t_j) b_i)(b_j) =$$

$$= \sum_{i,j=1}^{N} (F(J_{t_i} \cap J_{t_j}) b_i)(b_j) = \| \sum_{i=1}^{N} b_i 1_{J_{t_i}} \|^2_{\mathfrak{X}} .$$

Since $\mathfrak{X}$ is dense in $\overline{\mathfrak{X}}$ and the set of vectors $\sum_{i=1}^{N} X_{t_i} b_i$ is dense in H_X it follows that the map

$$U : \sum_{i=1}^{N} b_i 1_{J_{t_i}} \longrightarrow \sum_{i=1}^{N} X_{t_i} b_i$$

is an isometric isomorphism, which completes the proof of the lemma.

Further we use an B^* valued integral with respect to $L(B,B^*)$ valued positive additive measure F.

Let $f \in \mathfrak{X}$ and $f(t) = \sum_{i=1}^{N} b_i 1_{D_i}(t)$, where $D_i \in \mathfrak{R}$ and $b_i \in B$.
For every $J_t \in \mathfrak{R}$ we put

$$\int_{J_t} f(s) F(ds) = \sum_{i=1}^{N} F(D_i \cap J_t) b_i .$$

The B^* valued integral defined by the last formula does not depend on the choice of b_i and D_i in the representation of f. Moreover it is linear in f. In the sequel <u>we shall assume that the space</u> H_X <u>is separable</u>. In this case the space $\overline{\mathfrak{X}}$ is separable too.
Let $e_1, e_2, \dots$ be a CONS in $\overline{\mathfrak{X}}$ such that $e_1, e_2, \dots \in \mathfrak{X}$.
Putting

$$g_n = U e_n , \qquad n = 1,2,\dots ,$$

where U is the isometric isomorphism between $\overline{\mathfrak{X}}$ and H_X, as in the lemma 2.1, we obtain that $g_1, g_2, \dots$ is a CONS in H_X. For any $t \in T$

and $b \in B$ we have

$$X_t b = \sum_{n=1}^{\infty} g_n(X_t b, g_n)_{H_X} = \sum_{n=1}^{\infty} g_n(b 1_{J_t}, e_n)_{\mathfrak{X}} =$$

$$= \sum_{n=1}^{\infty} g_n[\sum_{k=1}^{N_n} (F(J_t \cap D_k^n) b_k^n)](b) = \sum_{n=1}^{\infty} g_n(\int_{J_t} e_n(s) F(ds))(b).$$

Thus we have proved the following Theorem which generalizes the well known Wiener expansion for Brownian motion.

Theorem 2.2. Every B-valued m-correlated process $(X_t)_{t \in T}$ over the semilattice T with the separable vector time domain H_X admits a generalized Wiener expansion

$$X_t b = \sum_{n=1}^{\infty} g_n(\int_{J_t} e_n(s) F(ds)) b,$$

for any $b \in B$ and $t \in T$, where the series is convergent in the norm of the space H_X.

Example 2.2. Let $T = R_+^d$ be a semilattice of d-vectors with positive real components and with the natural ordering. Consider a Brownian motion ξ_t with d-dimensional time.
Since its correlation functions is of the form

$$K(t,s) = E\, \xi_t\, \xi_s = \prod_{i=1}^{d} (t_i \wedge s_i) = K(t \wedge s),$$

thus ξ_t is m-correlated over semilattice R_+^d.

If we put $T = R_+^d$, $B = R$ and $F = dt$ - the Lebesque measure on T, then $\overline{\mathfrak{X}} = L_2(R_+^d, dt)$. Thus by Theorem 2.2 we obtain that

$$\xi_t = \sum_{n=1}^{\infty} \varphi_n \int_0^t e_n(s) ds, \qquad t \in R_+^d ,$$

where φ_n is a sequence of independent random variables with distribution $N(0,1)$, e_n is a CONS in $L_2(R_+^d, dt)$ and the series is convergent in L_2 (Ω,P). Moreover in this case one may prove that the series is convergent with probability 1, uniformly on every compact subset of R_+^d, cf. [4].

References

[1] S.A.Chobanyan,A.Weron, Banach space valued stationary processes and their linear prediction, Dissertationes Math. 125(1975), 1-45.

[2] R.K.Getoor, The shift operator for non-stationary stochastic processes, Duke Math. J. 23(1956), 175-187.

[3] H.Hamburger, Über eine Erweterung das Stieltjessehen Momentenproblems, Math.Annalen 81 (1920), 235-319, ibidem 82 (1921), 120-187.

[4] K.Ito,M.Nisio, On the convergence of sums of independent Banach space valued random variables, Osaka J.Math. 5 (1968), 35-48.

[5] Nguyen Van Thu, On additively correlated random variables, Bull.Acad.Polon. 23 (1975), 781-785.

[6] Nguyen Van Thu, Gaussian Markov processes on partially ordered sets, Comm.Math. (to appear).

[7] A.E.Nussbaum, Integral representation of semigroups of unbounded self-adjoint operators, Annales of Math. 69 (1959), 133-141.

[8] A.Weron, Prediction theory in Banach spaces, Lecture Notes in Math. vol.472 (1975), 207-228.

Institute of Mathematics
Wroclaw Technical University
Wybrzeże Wyspiańskiego 27
50-370 Wrocław
P o l a n d

PREDICTION THEORY FOR NON-STATIONARY SEQUENCES OF RANDOM VECTORS

Hannu Niemi

0. Introduction

We consider non-stationary purely non-deterministic sequences of random (column) vectors,

$$\underline{\underline{x}}_n = (x_n^{(1)}, \ldots, x_n^{(q)})^t, \qquad n \in Z,$$

for which there exists a polynomial P such that

$$(*) \qquad E|x_n^{(j)}|^2 \leq P(n) \quad \text{for all} \quad j = 1, \ldots, q; \quad n \in Z.$$

As Deo [3] (see also Dudley [4]) has noticed one can define a so-called spectral covariance distribution of any sequence of random vectors $\underline{\underline{x}}_n$, $n \in Z$, satisfying the condition (*).

We continue the work started by Dudley [4] and Deo [3]. We introduce a so-called maximal canonical representation for the spectral covariance distribution $\underline{\underline{B}}$ of a purely non-deterministic sequence of random vectors $\underline{\underline{x}}_n$, $n \in Z$, satisfying (*); and show that to a maximal canonical representation for $\underline{\underline{B}}$ it always corresponds a so-called Cramér-Hida representation for $\underline{\underline{x}}_n$, $n \in Z$. An analogous result concerning purely non-deterministic stationary sequences of random vectors has been presented by Rozanov [6; see pp. 61-62 and the references on pp. 200-201] (see also Wiener and Masani [8] and Masani [5]).

If $\underline{\underline{x}}_n$, $n \in Z$, is a sequence of random vectors satisfying (*), then even the sequence $\hat{\underline{\underline{x}}}_{n,(p)}$, $n \in Z$, formed by the best p-step linear least square predictions of $\underline{\underline{x}}_n$, $n \in Z$, satisfies (*). In section 4 we present a representation for the spectral covariance distribution $\hat{\underline{\underline{B}}}_{(p)}$ of $\hat{\underline{\underline{x}}}_{n,(p)}$, $n \in Z$, in terms of maximal canonical representation for the spectral covariance distribution $\underline{\underline{B}}$ of $\underline{\underline{x}}_n$, $n \in Z$.

Finally, it is presented a connection between a maximal canonical representation for the spectral covariance distribution $\underline{\underline{B}}$ of a purely non-deterministic stationary sequence of random vectors $\underline{\underline{x}}_n$, $n \in Z$, and a maximal matrix related to the spectral density matrix of $\underline{\underline{x}}_n$, $n \in Z$ (see Rozanov [6; p. 60]).

1. The Cramér-Hida representation

Let $(\Omega, \mathcal{A}, P)$ be a probability space. By $L_0^2(\Omega, \mathcal{A}, P)$ we denote the (complex) Hilbert space of all complex-valued random variables defined on $(\Omega, \mathcal{A}, P)$ with zero mean and finite second order moments.

In this paper we consider sequences of random (column) vectors

$$\underline{\underline{x}}_n = (x_n^{(1)}, \ldots, x_n^{(q)})^t, \qquad n \in Z,$$

for which $x_n^{(j)} \in L_0^2(\Omega, \mathcal{A}, P)$ for all $j = 1, \ldots, q$; $n \in Z$.

Let $\underline{\underline{x}}_n$, $n \in Z$, be a sequence of random vectors. For $k \in Z$ by $\overline{sp}\{\underline{\underline{x}};k\}$ we denote the closed linear subspace in $L_0^2(\Omega, \mathcal{A}, P)$ spanned by the set $\{x_n^{(j)} \mid n \leq k,\ j = 1, \ldots, q\}$; and by $\overline{sp}\{\underline{\underline{x}}\}$ we denote the closed linear subspace in $L_0^2(\Omega, \mathcal{A}, P)$ spanned by the set $\{x_n^{(j)} \mid n \in Z,\ j = 1, \ldots, q\}$. Furthermore, we use the notation

$$\overline{sp}\{\underline{\underline{x}}; -\infty\} = \bigcap_{n \in Z} \overline{sp}\{\underline{\underline{x}};n\}.$$

A sequence of random vectors $\underline{\underline{x}}_n$, $n \in Z$, is called _purely non-deterministic_, if

$$\overline{sp}\{\underline{\underline{x}}; -\infty\} = \{0\};$$

it is called _deterministic_, if

$$\overline{sp}\{\underline{\underline{x}}; -\infty\} = \overline{sp}\{\underline{\underline{x}}\}.$$

We recall that any sequence of random vectors can be represented as the sum of a deterministic and a purely non-deterministic component (Cramér [2; Theorem 1]).

Let H be an arbitrary Hilbert space and let $A \subset H$. By $\overline{sp}\{A\}$ we denote the closed linear subspace in H spanned by A. Let $M \subset H$ be a closed linear subspace in H. By P_M we denote the orthogonal projection of H onto M.

Let $\underline{\underline{x}}_n$, $n \in Z$, be a purely non-deterministic sequence of random vectors. For $n \in Z$ put $P_n = P_{\overline{sp}\{\underline{\underline{x}};n\}}$; and denote $P_{n-1}\,\underline{\underline{x}}_n = (P_{n-1}x_n^{(1)},\ldots,P_{n-1}x_n^{(q)})^t$,

$$\underline{\underline{\xi}}_n = \underline{\underline{x}}_n - P_{n-1}\underline{\underline{x}}_n, \qquad n \in Z.$$

Furthermore, put

(1) $$\underline{\underline{\eta}}_n = (\eta_n^{(1)},\ldots,\eta_n^{(s_n)})^t,$$

where $\eta_n^{(1)},\ldots,\eta_n^{(s_n)}$, is a (fixed) orthonormal basis in the linear subspace in $\overline{sp}\{\underline{\underline{x}}\}$ spanned by the vectors $\xi_n^{(1)},\ldots,\xi_n^{(q)}$. (Put $\underline{\underline{\eta}}_n = 0$, if $\xi_n^{(1)} = \ldots = \xi_n^{(q)} = 0$.)

For each $m \in Z$, the number $r(\underline{\underline{x}};m) = s_m$ is called the <u>rank of the innovation of</u> $\underline{\underline{x}}_n$, $n \in Z$, <u>at</u> m.

The following representation theorem is due to Cramér [2; Theorem 1]. We state it as presented by Deo [3; p. 127].

<u>Theorem 1</u> (Cramér). Let $\underline{\underline{x}}_n$, $n \in Z$, <u>be a purely non-deterministic sequence of random vectors. Then there exist</u> $q \times s_n$ <u>matrices</u> $(s_n = r(\underline{\underline{x}};n))$

$$\underline{\underline{A}}_{n,p} = (a_{n,p}^{(hj)})_{q \times s_n}, \qquad p \le n;$$

$n \in Z$, <u>such that</u>

(2) $$x_n^{(j)} = \sum_{k=-\infty}^{n} \sum_{s=1}^{s_k} a_{n,k}^{(js)}\,\eta_k^{(s)},$$

$$j = 1,\ldots,q; \quad n \in Z.$$

(<u>The series</u> (2) <u>is convergent in</u> $L_0^2(\Omega,\mathcal{A},P)$.)

2. On sequences of random vectors satisfying a polynomial growth condition

In this section we recall a characterization of purely non-deterministic sequences of random vectors $\underline{\underline{x}}_n$, $n \in Z$, satisfying the growth condition:

(3) $$E|x_n^{(j)}|^2 \le P(n), \qquad j = 1,\ldots,q; \quad n \in Z,$$

for some polynomial P. The characterization is essentially due to Deo [3; Theorem 4] (see also Dudley [4; Theorem 1]).

For the sake of clarity, we give a complete proof of the characterization.

In what follows we use some results concerning Fourier coefficients of (Schwartz) distributions on the k-dimensional torus T^k, i.e., on the Cartesian product of k unit circles T in the complex plane; $k = 1, 2$. (For the definition see Schwartz [7; pp. 224–227].)

Let $\underline{\underline{x}}_n$, $n \in Z$, be a sequence of random vectors satisfying the growth condition (3). For any $h, j = 1, \ldots, q$ there then exists a (uniquely determined) distribution $B^{(hj)}$ on T^2 for which

$$B^{(hj)}(\chi_m \otimes \overline{\chi}_n) = E\, x_m^{(h)}\, \overline{x_n^{(j)}}, \qquad m, n \in Z;$$

here $\chi_m(s) = \exp(-2\pi i m s)$, $s \in [0,1]$, and $\chi_m \otimes \overline{\chi}_n\,(s,t) =$

$$= \chi_m(s)\, \overline{\chi_n(t)}\,;\ s,\ t \in [0,1].$$

The matrix

$$\underline{\underline{B}} = (B^{(hj)})_{q \times q}$$

of distribution on T^2 is called the <u>spectral covariance distribution</u> of the sequence $\underline{\underline{x}}_n$, $n \in Z$.

According to Deo [3] we call a $q \times r$ matrix of distributions a <u>matrix valued distribution</u>. A matrix valued distribution $\underline{\underline{U}} = (U^{(hj)})_{q \times r}$, defined on T, is called <u>zero-meromorphic</u>, if there exists a representation

$$U^{(hj)} = \sum_{k=n'}^{\infty} a_k^{(hj)}\, \overline{\chi}_k$$

($a_{n'}^{(hj)} \neq 0$ for some pair (hj)) for all $h = 1,\ldots,q$; $j = 1,\ldots,r$, i.e., the Fourier coefficients $a_n^{(hj)} = U^{(hj)}(\chi_n)$, $n \in Z$, are equal to zero for all integers $n < n'$.

Since the sequence $a_n^{(hj)}$, $n \geq n'$, is of polynomial growth as $n \longrightarrow \infty$, the series

$$\Phi^{(hj)}(z) = \sum_{k=n'}^{\infty} a_k^{(hj)}\, z^k \tag{4}$$

is the Laurent series of a function analytic on the unit disc $|z| < 1$; expect possibly for a pole at 0 (cf. Dudley [4;p.228]).

Let $\underline{\underline{U}} = (U^{(hj)})_{q \times r}$ be a matrix valued distribution on T. Then the matrix valued distribution

$$\underline{\underline{U}} \otimes \underline{\underline{U}}^* = (A^{(hj)})_{q \times q}$$

on T^2 is defined by

(5) $$A^{(hj)} = \sum_{s=1}^{r} U^{(hs)} \otimes \overline{U}^{(js)}, \qquad h,j = 1,\dots,q.$$

<u>Remark</u>. Let $\underline{\underline{B}}$ be a matrix valued distribution on T^2. By writing

(6) $$\underline{\underline{B}} = \sum_{k=-\infty}^{\infty} \underline{\underline{U}}_k \otimes \underline{\underline{U}}_k^*,$$

where $\underline{\underline{U}}_n = (U_n^{(hj)})_{q \times r}$, $n \in Z$, are matrix valued distributions on T we mean that every component in the series (6) is convergent in the weak* topology of all distributions on T^2. Since the space of all distributions on T^2 is a Montel space, this implies that each component in the series (6) is convergent in the strong topology of all the distributions on T^2 (see Dudley [4; p. 225] or Abreu [1; pp. 391-397]).

The following lemma is essentially due to Dudley [4; pp. 229-230]. Before presenting the lemma we note that the spectral covariance distribution $\underline{\underline{B}}$ of any sequence of random vectors $\underline{\underline{x}}_n$, $n \in Z$, satisfying the condition (3) can be presented as

$$\underline{\underline{B}} = \sum_{k=-\infty}^{\infty} \underline{\underline{U}}_k \otimes \underline{\underline{U}}_k^*,$$

where $\underline{\underline{U}}_n = (U_n^{(hj)})_{q \times r_n}$, $n \in Z$, is a matrix valued distribution on T (cf. Deo [3; Theorem 3]).

<u>Lemma 2.</u> <u>Let</u> $\underline{\underline{x}}_n$, $n \in Z$, <u>be a sequence of random vectors satisfying the growth condition</u> (3). <u>Suppose</u>

$$\underline{\underline{B}} = \sum_{k=-\infty}^{\infty} \underline{\underline{U}}_k \otimes \underline{\underline{U}}_k^* ;$$

<u>where</u> $\underline{\underline{U}}_n = (U_n^{(hj)})_{q \times r_n}$, $n \in Z$, <u>is a matrix valued distribution on</u> T, <u>is a representation for the spectral covariance distribution</u> $\underline{\underline{B}}$ <u>of</u> $\underline{\underline{x}}_n$, $n \in Z$. <u>If</u> H <u>is any separable infinite dimensional complex Hilbert space and if</u> $e_n^{(s)}$, $s = 1,\dots,r_n$, $n \in Z$, <u>is (a renumbering of) an orthonormal basis in</u> H, <u>then for every</u> $j = 1,\dots,q$; $n \in Z$, <u>the series</u>

(7) $$y_n^{(j)} = \sum_{k=-\infty}^{\infty} \sum_{s=1}^{r_k} U_k^{(js)}(\chi_n) e_k^{(s)}$$

<u>is convergent in</u> H <u>and the inner product of any</u> $y_m^{(h)}$ and $y_n^{(j)}$ <u>is</u>

$$(8) \qquad (y_m^{(h)} \mid y_n^{(j)}) = \sum_{k=-\infty}^{\infty} \sum_{s=1}^{r_k} U_k^{(hs)}(\chi_m)\,\overline{U}_k^{(js)}(\chi_n) =$$

$$= B^{(hj)}(\chi_m \otimes \overline{\chi}_n) = E\, x_m^{(h)}\, \overline{x_n^{(j)}} ,$$

$$h, j = 1, \ldots, q;\ m, n \in Z.$$

<u>Remark.</u> Let $\underline{\underline{x}}_n$, $n \in Z$, be a sequence of random vectors satisfying (3). Then, with the notation of Lemma 2, there exists an inner product preserving isomorphism $L:\ \overline{sp}\{\underline{\underline{x}}\} \longrightarrow \overline{sp}\{y_n^{(j)} \mid j = 1, \ldots, q;\ n \in Z\}$ such that

$$Lx_n^{(j)} = y_n^{(j)} \quad \text{for all} \quad j = 1, \ldots, q;\ n \in Z.$$

<u>Definition 3.</u> <u>The innovation rank</u> $r(\underline{\underline{U}})$ of a zero-meromorphic matrix valued distribution $\underline{\underline{U}} = (U^{(hj)})_{q \times r}$;

$$U^{(hj)} = \sum_{k=n'}^{\infty} a_k^{(hj)}\, \overline{\chi}_k, \quad h = 1, \ldots, q;\ j = 1, \ldots, r,$$

($a_{n'}^{(hj)} \neq 0$ for some pair (hj)) is equal to the rank of the matrix

$$\underline{\underline{U}}(\chi_{n'}) = (U^{(hj)}(\chi_{n'}))_{q \times r} .$$

We are now ready to present the characterization of a purely non-deterministic sequence of random vectors satisfying (3).

<u>Theorem 4</u> (Deo, Dudley). <u>A sequence of random vectors</u> $\underline{\underline{x}}_n$, $n \in Z$, <u>satisfying the growth condition</u> (3) <u>is purely non-deterministic, if and only if the spectral covariance distribution</u> $\underline{\underline{B}} = (B^{(hj)})_{q \times q}$ <u>of</u> $\underline{\underline{x}}_n$, $n \in Z$, <u>can be represented as</u>

$$(9a) \qquad \underline{\underline{B}} = \sum_{k=-\infty}^{\infty} \underline{\underline{U}}_k \otimes \underline{\underline{U}}_k^* ;$$

<u>where for all</u> $n \in Z$: $\underline{\underline{U}}_n = (U_n^{(hj)})_{q \times r_n}$ <u>is a zero-meromorphic matrix valued distribution on</u> T <u>such that</u>

(9b) $$U_n^{(hj)} = \sum_{k=m(n)}^{\infty} c_{k,n}^{(hj)} \overline{\chi_k} ,$$

(9c) $$c_{m(n),n}^{(hj)} \neq 0 \quad \underline{\text{for some pair}} \quad (hj).$$

If $\underline{\underline{x}}_n$, $n \in Z$, is a purely non-deterministic sequence of random vectors satisfying (3), then there exists such a representation for $\underline{\underline{B}}$ in the form (9a) that for all $n \in Z$

(i) $m(n) < m(n+1)$

(ii) $r_n = r(\underline{\underline{U}}_n) = r(\underline{\underline{x}};m(n))$ and

(iii) $r(\underline{\underline{x}};k) \neq 0$, if and only if there exists (exactly) one integer j for which $m(j) = k$.

Proof. Suppose $\underline{\underline{x}}_n$, $n \in Z$, is a purely non-deterministic sequence of random vectors satisfying (3). As Deo [3; pp.127-128] (see also Dudley [4; pp. 228-229]) has shown, for each $n \in Z$, the matrix valued distribution $\underline{\underline{U}}_n = (U_n^{(hj)})_{q \times r_n}$ can be defined as

$$U_n^{(hj)} = \sum_{k=n}^{\infty} a_{k,n}^{(hj)} \overline{\chi_k} ;$$

where the matrices $\underline{\underline{A}}_{m,p} = (a_{m,p}^{(hj)})_{q \times r_m}$, $m \in Z$, $p \leq m$, are defined as in (2).

The conditions (i) - (iii) (and the condition $c_{m(n),n}^{(hj)} \neq 0$ for some pair (hj)) are satisfied, if we omit all $\underline{\underline{U}}_n$ corresponding to an index n for which $r(\underline{\underline{x}};n) = 0$; and make an obvious renumbering of the (remaining) elements in the sequence $\underline{\underline{U}}_n$, $n \in Z$.

The latter part of the proof is based on Lemma 2 (cf.Dudley [4; pp. 229-230]).

Suppose $\underline{\underline{x}}_n$, $n \in Z$, is a sequence of random vectors satisfying (3). Suppose the spectral covariance distribution $\underline{\underline{B}}$ of $\underline{\underline{x}}_n$, $n \in Z$, can be represented in the form (9a-c).

Let H be any separable infinite dimensional complex Hilbert space and let $e_n^{(s)}$, $s = 1,\ldots,r_n$; $n \in Z$, be (a renumbering of) an orthonormal basis in H. It then follows from Lemma 2 that for the elements

$$y_n^{(j)} = \sum_{k=-\infty}^{\infty} \sum_{s=1}^{r_k} U_k^{(js)}(\chi_n)\, e_k^{(s)}, \quad j = 1,\dots,q;\ n \in Z,$$

we have

$$(y_m^{(h)} \mid y_n^{(j)}) = E\, x_m^{(h)}\, \overline{x_n^{(j)}}, \quad h,j = 1,\dots,q;\ m,\ n \in Z.$$

For $n \in Z$ put $\overline{sp}\{\underline{y};n\} = \overline{sp}\,\{y_k^{(j)} \mid j = 1,\dots,q;\ k \leq n\}$; and

$$\overline{sp}\{\underline{y};-\infty\} = \bigcap_{n \in Z} \overline{sp}\{\underline{y};n\}.$$

We show that $\overline{sp}\{\underline{y};-\infty\} = \{0\}$. Choose a vector $e_{k'}^{(s')}$. By assumption all $\underline{\underline{U}}_n$, $n \in Z$, are zero-meromorphic; thus there exists an integer $n(s')$ such that

$$(y_n^{(j)} \mid e_{k'}^{(s')}) = 0 \quad \text{for all} \quad j = 1,\dots,q,$$

when $n < n(s')$. Therefore $e_{k'}^{(s')} \perp \overline{sp}\{\underline{y};n\}$ when $n < n(s')$; and a fortiori $e_{k'}^{(s')} \perp \overline{sp}\{\underline{y};-\infty\}$. This implies that $\overline{sp}\{\underline{y};-\infty\} = \{0\}$.

Since $(y_m^{(h)} \mid y_n^{(j)}) = E\, x_m^{(h)}\, \overline{x_n^{(j)}}$ for all $h,j = 1,\dots,q;m,$ $n \in Z$, it then follows that even $\overline{sp}\{\underline{\underline{x}};-\infty\} = \{0\}$.

The theorem is proved.

Definition 5. Let $\underline{\underline{x}}_n$, $n \in Z$, be a purely non-deterministic sequence of random vectors satisfying the growth condition (3). A representation (9a-c) for the spectral covariance distribution $\underline{\underline{B}}$ of $\underline{\underline{x}}_n$, $n \in Z$, is canonical, if the conditions (i) - (iii) stated in Theorem 4 are satisfied.

Remark. Let $\underline{\underline{x}}_n$, $n \in Z$, be a purely non-deterministic sequence of random vectors satisfying the growth condition (3). Suppose there exists two canonical representations

$$\underline{\underline{B}} = \sum_{k=-\infty}^{\infty} \underline{\underline{U}}_k \otimes \underline{\underline{U}}_k^* = \sum_{k=-\infty}^{\infty} \underline{\underline{V}}_k \otimes \underline{\underline{V}}_k^*$$

for the spectral covariance distribution $\underline{\underline{B}}$ of $\underline{\underline{x}}_n$, $n \in Z$; where

$$\underline{\underline{U}}_n = (U_n^{(hj)})_{q \times r_n}, \quad \underline{\underline{V}}_n = (V_n^{(hj)})_{q \times r'_n},$$

$$U_n^{(hj)} = \sum_{k=m(n)}^{\infty} c_{k,m}^{(hj)} \overline{\chi}_k, \quad V_n^{(hj)} = \sum_{k=m'(n)}^{\infty} b_{k,n}^{(hj)} \overline{\chi}_k$$

($c_{m(n),n}^{(hj)} \neq 0$ (resp. $b_{m'(n),n}^{(hj)} \neq 0$) for some pair (hj)). Then there exists a uniquely determined integer p for which

$$m(n+p) = m'(n) \qquad \text{for all} \quad n \in Z;$$

and a fortiori

$$r_{n+p} = r'_n \qquad \text{for all } n \in Z.$$

The following definition is an analogue of the definition of a maximal matrix related to the spectral density matrix of a stationary purely non-deterministic sequence of random vectors $\underline{\underline{x}}_n$, $n \in Z$(see Rozanov [6; p. 60 and the references given on pp. 200–201]).

Definition 6. Let $\underline{\underline{x}}_n$, $n \in Z$, be a purely non-deterministic sequence of random vectors satisfying the growth condition (3). A canonical representation for the spectral covariance distribution $\underline{\underline{B}}$ of $\underline{\underline{x}}_n$, $n \in Z$, in the form (9a–c) is maximal, if for any other canonical representation

$$\underline{\underline{B}} = \sum_{k=-\infty}^{\infty} \underline{\underline{V}}_k \otimes \underline{\underline{V}}_k^* ,$$

where $\underline{\underline{V}}_n = (V_n^{(hj)})_{q \times r'_n}$,

$$V_n^{(hj)} = \sum_{k=m'(n)}^{\infty} b_{k,n}^{(hj)} \overline{\chi}_k$$

($b_{m'(n),n}^{(hj)} \neq 0$ for some pair (hj)), $n \in Z$, the following condition is satisfied:

Suppose p is such an integer that $m'(n) = m(n+p)$ for all $n \in Z$. Then

$$(10) \quad \sum_{h=1}^{q} \sum_{j=1}^{q} a_h \overline{a}_j \Big(\sum_{s=1}^{r_{n+p}} \Phi_{n+p}^{(hs)}(0) \overline{\Phi_{n+p}^{(js)}}(0) - \sum_{s=1}^{r'_n} \Psi_n^{(hs)}(0) \overline{\Psi_n^{(js)}}(0) \Big) \geq 0$$

for all $a_h \in C$, $h = 1,\dots,q$; here

$$\Phi_n^{(hj)}(z) = \sum_{k=m(n)}^{\infty} c_{k,m}^{(hj)} \; z^{k-m(n)} ,$$

$$\Psi_n^{(hj)}(z) = \sum_{k=m'(n)}^{\infty} b_{k,n}^{(hj)} \; z^{k-m'(n)} \; ; \quad |z| < 1.$$

Remark. There exist several maximal canonical representations for the spectral covariance distribution $\underline{\underline{B}}$ of a sequence of random vectors $\underline{\underline{x}}_n$, $n \in Z$, satisfying the growth condition (3).

The proof of the following theorem is essentially based on the same technique as used by Kozanov [6; pp. 60-61] in proving a corresponding theorem for stationary purely non-deterministic sequences of random vectors $\underline{\underline{x}}_n$, $n \in Z$.

Theorem 7. *Let* $\underline{\underline{x}}_n$, $n \in Z$, *be a purely non-deterministic sequence of random vectors satisfying the condition* (3). *Then a canonical representation for the spectral covariance distribution* $\underline{\underline{B}}$ *of* $\underline{\underline{x}}_n$, $n \in Z$, *in the form* (9a-c) *is maximal, if and only if for any (renumbering of an) orthonormal basis* $e_n^{(s)}$, $s = 1,\dots,r_n$; $n \in Z$, *in a separable infinite dimensional complex Hilbert space* H *the sequence*

$$y_n^{(j)} = \sum_{k \in I_n} \sum_{s=1}^{r_k} U_k^{(js)} (\chi_n) \, e_k^{(s)}, \quad j = 1,\dots,q; \quad n \in Z;$$

where $I_m = \{k \mid m(k) \leq m\}$, $m \in Z$, *has the property: For all* $n \in Z$ *has the property* :

$$\overline{sp}\{y_m^{(j)} \mid j = 1,\dots,q;\ m \leq n\} = \overline{sp}\{e_k^{(s)} \mid s = 1,\dots,r_k;\ k \in I_n\} .$$

If

$$\underline{\underline{B}} = \sum_{k=-\infty}^{\infty} \underline{\underline{U}}_k \otimes \underline{\underline{U}}_k^*$$

is a maximal canonical representation for $\underline{\underline{B}}$ *in the form* (9a-c), *then there exists a Cramér-Hida representation for* $\underline{\underline{x}}_n$, $n \in Z$,

<u>in the form</u>

$$x_n^{(j)} = \sum_{k\in I_n} \sum_{s=1}^{r_k} U_k^{(js)}(\chi_n)\, \tilde{\eta}_k^{(s)}, \quad j = 1,\ldots,q; \quad n \in Z,$$

<u>that is, for all</u> j <u>and</u> n <u>satisfying</u> $m(n) = j$ <u>the vectors</u> $\tilde{\eta}_n^{(s)} \in \overline{sp}\{\underline{\underline{x}}\}$ <u>span the same (closed) linear subspace in</u> $\overline{sp}\{\underline{\underline{x}}\}$ <u>as the vectors</u> $\eta_j^{(s)}$, $s = 1,\ldots,s_j$ $(= r_n)$, <u>defined in</u> (1).

<u>Proof.</u> Suppose a representation

(11)
$$\underline{\underline{B}} = \sum_{k=-\infty}^{\infty} \underline{\underline{U}}_k \otimes \underline{\underline{U}}_k^*,$$

where, for all $n \in Z$, $\underline{\underline{U}}_n = (U_n^{(hj)})_{q\times r_n}$,

$$U_n^{(hj)} = \sum_{k=m(n)}^{\infty} c_{k,n}^{(hj)}\, \overline{\chi}_k$$

$(c_{m(n),n}^{(hj)} \neq 0$ for some pair $(hj))$, is a (canonical) representation for the spectral covariance distribution $\underline{\underline{B}}$ of $\underline{\underline{x}}_n$, $n \in Z$, corresponding to a Cramér-Hida representation

(12)
$$x_n^{(j)} = \sum_{k\in I_n} \sum_{s=1}^{r_k} c_{n,k}^{(sj)}\, \eta_k^{(s)}, \quad j = 1,\ldots,q; \quad n \in Z,$$

in the form (2); here $I_m = \{k \mid m(k) \leq m\}$, $m \in Z$ (cf. the first part of the proof of Theorem 4). (Note that the condition stated in the theorem is then satisfied for any sequence

$$y_n^{(j)} = \sum_{k\in I_n} \sum_{s=1}^{r_k} U_k^{(js)}(\chi_n)\, e_k^{(s)}, \quad j = 1,\ldots,q; \quad n \in Z.)$$

We show that the representation (11) is maximal.

Suppose

$$\underline{\underline{B}} = \sum_{k=-\infty}^{\infty} \underline{\underline{V}}_k \otimes \underline{\underline{V}}_k^*,$$

where $\underline{\underline{V}}_n = (V_n^{(hj)})_{q\times r_n'}$,

$$V_n^{(hj)} = \sum_{k=m'(n)}^{\infty} b_{k,n}^{(hj)} \overline{\chi}_k$$

$(b_{m'(n),n}^{(hj)} \neq 0$ for some pair $(hj))$, $n \in Z$, is a canonical representation for $\underline{\underline{B}}$.

Suppose H is a separable infinite dimensional complex Hilbert space and suppose $f_n^{(s)}$, $s = 1,\ldots,r_n'$; $n \in Z$, is (a renumbering of) an orthonormal basis in H. Consider the sequence

$$z_n^{(j)} = \sum_{k \in I_n'} \sum_{s=1}^{r_k} V_k^{(js)}(\chi_n)\, f_k^{(s)}, \quad j = 1,\ldots,q;\ n \in Z;$$

where $I_m' = \{ k \mid m'(k) \leq m \}$, $m \in Z$. For $n \in Z$ put

$$\overline{sp}\{\underline{\underline{z}};n\} = \overline{sp}\{z_k^{(j)} \mid j = 1,\ldots,q;\ k \leq n \},$$

$$\overline{sp}\{\underline{\underline{f}};n\} = \overline{sp}\{f_k^{(s)} \mid s = 1,\ldots,r_k';\ k \in I_n' \}.$$

Clearly,

$$\overline{sp}\{\underline{\underline{z}};n\} \subset \overline{sp}\{\underline{\underline{f}};n\} \quad \text{for all} \quad n \in Z.$$

Suppose $p \in Z$ is such an integer that $m(n+p) = m'(n)$ for all $n \in Z$. Since (12) is a Cramér-Hida reprezentation for $\underline{\underline{x}}_n$, $n \in Z$, it then follows that for any vectors

$$u = \sum_{s=1}^{q} a_s\, x_j^{(s)} \quad \text{and} \quad v = \sum_{s=1}^{q} a_s\, z_j^{(s)} ,$$

where $a_s \in C$, $s = 1,\ldots,q$, and $j \in Z$ we have

$$\| P_{\overline{sp}\{\underline{\underline{x}};j-1\}} u - u \|^2_{\overline{sp}\{\underline{\underline{x}}\}} = \| P_{\overline{sp}\{\underline{\underline{z}};j-1\}} v - v \|_H^2 \geqslant$$

$$\geqslant \| P_{\overline{sp}\{\underline{\underline{f}};j-1\}} v - v \|_H^2$$

or equivalently

$$(13)\qquad \sum_{=1}^{q}\sum_{k=1}^{q} a_n \bar{a}_k \sum_{s=1}^{r_{n+p}} c^{(sh)}_{n+p,m(n+p)} \overline{c^{(sk)}_{n+p,m(n+p)}} \geqslant$$

$$\geqslant \sum_{h=1}^{q}\sum_{k=1}^{q} a_n \bar{a}_k \sum_{s=1}^{r'_n} {}^{(sh)}_{n,m'(n)} \overline{b^{(sk)}_{n,m'(n)}} ;$$

where $n \in Z$ is the integer satisfying the condition $m(j) = n$. The inequality (13) is clearly equivalent to the condition (10).

On the other hand, suppose a canonical representation

$$(14)\qquad \underline{\underline{B}} = \sum_{k=-\infty}^{\infty} \underline{\underline{V}}_k \otimes \underline{\underline{V}}_k^{*},$$

where $\underline{\underline{V}}_n = (V_n^{(hj)})_{q \times r'_n}$,

$$V_n^{(hj)} = \sum_{k=m'(n)}^{\infty} b^{(hj)}_{k,n} \bar{\chi}_k$$

$(b^{(hj)}_{m'(n),n} \neq 0$ for some pair $(hj))$, $n \in Z$, is maximal.

Let H be a separable infinite dimensional complex Hilbert space and let $e_n^{(s)}$, $s = 1,\dots,r'_n$, $n \in Z$, be (a renumbering of) an orthonormal basis in H. Consider the sequence

$$(15)\qquad y_n^{(j)} = \sum_{k \in I'_n} \sum_{s=1}^{r'_k} V_k^{(hj)}(\chi_n)\, e_k^{(s)}, \qquad j = 1,\dots,q; \quad n \in Z,$$

where $I'_m = \{k \mid m'(k) \leq m\}, m\ Z$, and the Cramér-Hida representation (12) for $\underline{\underline{x}}_n$, $n \in Z$.

Let $L: \overline{sp}\{\underline{\underline{x}}\} \longrightarrow \overline{sp}\{y_n^{(j)} \mid j = 1,\dots,q;\ n \in Z\}$ be an inner product preserving isomorphism such that

$$Lx_n^{(j)} = y_n^{(j)} \quad \text{for all } j = 1,\dots,q; \quad n \in Z.$$

Put $f_n^{(s)} = L\eta_n^{(s)}$, $s = 1,\dots,r_n$; $n \in Z$. Then

$$y_n^{(j)} = \sum_{k \in I_n} \sum_{s=1}^{r_k} c_{n,k}^{(sj)} f_n^{(s)}, \quad j = 1,\ldots,q; \quad n \in Z;$$

and since (12) is a Cramér-Hida representation for $\underline{\underline{x}}_n$, $n \in Z$, we get

$$\overline{sp}\{y_m^{(j)} | j = 1,\ldots,q; \ m \leq n\} = \overline{sp}\{f_m^{(s)} | s = 1,\ldots,r_m; m \in I_n\}$$

for all $n \in Z$.

Let p be an integer for which $m(n+p) = m'(n)$ for all $n \in Z$. It then follows from the maximality of the representation (14) that

$$(16) \qquad \sum_{s=1}^{r_{n+p}} c_{n+p,m(n+p)}^{(sh)} \overline{c_{n+p,m(n+p)}^{(sh)}} \leq \sum_{s=1}^{r'_n} b_{n,m'(n)}^{(sh)} \overline{b_{n,m'(n)}^{(sh)}}$$

for all $h = 1,\ldots,q$; $n \in Z$.

For $n \in Z$ put

$$\overline{sp}\{\underline{y};n\} = \overline{sp}\{y_m^{(j)} | j = 1,\ldots,q; \ m \leq n\},$$

$$\overline{sp}\{\underline{e};n\} = \overline{sp}\{e_m^{(s)} | s = 1,\ldots,r'_m \quad m \in I'_n\}$$

Combininig the inequalities (13) and (16) we then get

$$\| y_j^{(h)} - P_{\overline{sp}\{\underline{y};j-1\}} y^{(h)} \|_H = \| y_j^{(h)} - P_{\overline{sp}\{\underline{e};j-1\}} y_j^{(h)} \|_H$$

for all integers j for which there exists an integer n such that $m'(n) = j$; and $h = 1,\ldots,q$. Furthermore, since

$$\overline{sp}\{\underline{y};j-1\} \subset \overline{sp}\{\underline{e};j-1\},$$

it follows from the projection theorem that

$$(17) \qquad y_j^{(h)} - P_{\overline{sp}\{\underline{y};j-1\}} y_j^{(h)} = y_j^{(h)} - P_{\overline{sp}\{\underline{e};j-1\}} y_j^{(h)}$$

for all $h = 1,\ldots,q$. But this means that the vectors $f_{n+p}^{(s)}$, $s = 1,\ldots,r_{n+p}$, and $e_n^{(s)}$, $s = 1,\ldots,r'_n$ $(= r_{n+p})$ span the same (closed) linear subspace in H. Since this is true for all $n \in Z$ we get, $\overline{sp}\{\underline{y};m\} = \overline{sp}\{\underline{e};m\}$ for all $m \in Z$.

The first part of the theorem is proved.

For the proof of the second part of the theorem we note that it follows from the first part of the proof that $e_n^{(s)} \in \overline{sp}\{y_k^{(j)} \mid j = 1,\ldots,q;\ k \in Z\}$ for all $s = 1,\ldots,r_n'$; $n \in Z$. Thus we can define

$$\tilde{\eta}_n^{(s)} = L^{-1}e_n^{(s)}, \quad s = 1,\ldots,r_n' ; \quad n \in Z,$$

and, a fortiori,

$$x_n^{(j)} = L^{-1}y_n^{(j)} = \sum_{k\in I_n'} \sum_{s=1}^{r_k'} V_k^{(js)}(\chi_n)\, \tilde{\eta}_k^{(s)}, \quad j = 1,\ldots,q; \quad n \in Z.$$

The theorem is proved.

4. Linear prediction

Let $\underline{x}_n$, $n \in Z$, be a sequence of random vectors. For any integer $p > 0$ the best linear least square prediction of the component $x_n^{(j)}$ in terms of all the variables $x_k^{(1)},\ldots,x_k^{(q)}$ with $k \leq n-p$ is, by definition, the projection $P_{n-p}x_n^{(j)}$, $j = 1,\ldots,q$: $n \in Z$.

The following theorem is due to Cramér [2; Theorem 2].

__Theorem 8__ (Cramér). __Let__ $\underline{x}_n$, $n \in Z$, __be a purely non-deterministic sequence of random vectors. Then, with the notation of Theorem__ 1, __for any integers__ $p > 0$, $n \in Z$ __and__ $j = 1,\ldots,q$

$$P_{n-p}\, x_n^{(j)} = \sum_{k=-\infty}^{n-p} \sum_{s=1}^{s_k} a_{n,k}^{(js)}\ \eta_k^{(s)} .$$

__The prediction error is__

$$\sigma_{n,p}^{(j)} = \{E\,|x_n^{(j)} - P_{n-p}\, x_n^{(j)}|^2\}^{1/2} =$$

$$= \{\sum_{k=n-p+1}^{n} b_{n,k}^{(j)}\}^{1/2} ,$$

__where__

$$b_{n,k}^{(j)} = \sum_{s=1}^{s_k} | a_{n,k}^{(js)} |^2 , \quad j = 1,\dots,q; \quad n \in Z.$$

In this section we present another approach to the linear prediction problem by applying Theorem 7.

In order to clarify our method we consider first an elementary example.

Example 9. Let U be a zero-meromorphic (scalar) distribution on T and let

$$a_n = U(\chi_n), \quad n \in Z .$$

Then

$$U = \sum_{k=n'}^{\infty} a_k \overline{\chi_k} ,$$

where n' is the integer satisfying the conditions: $a_{n'} \neq 0$ and $a_n = 0$ for all $n < n'$.

Fix an integer $p > 0$. Then, by using the notation of linear prediction,

$$P_{n-p} a_n = 0 \qquad \text{if} \quad n < n' + p; \quad \text{and}$$

$$P_{n-p} a_n = U_{(p)}(\chi_n) \quad \text{if} \quad n \geq n' + p,$$

where

$$U_{(p)} = \chi_p \sum_{k=n'}^{\infty} a_k \overline{\chi_k} .$$

Thus,

$$P_{n-p} a_n = \hat{U}_{(p)}(\chi_n) \qquad \text{for all} \quad n \in Z,$$

if

$$\hat{U}_{(p)} = \chi_p \sum_{k=n'}^{\infty} a_{k+p} \overline{\chi_{k+p}} . \tag{18}$$

For the prediction error we get

$$\sigma_{n,p} = a_n , \qquad \text{if} \quad n' \leq n < n' + p; \quad \text{and}$$

$$\sigma_{n,p} = 0 \qquad \text{in the other cases.}$$

Theorem 10. Let $\underline{\underline{x}}_n$, $n \in Z$, be a purely non-deterministic sequence of random vectors satisfying the growth condition (3). Let

$$(19) \qquad \underline{\underline{B}} = \sum_{k=-\infty}^{\infty} \underline{\underline{U}}_k \otimes \underline{\underline{U}}_k^*$$

be a maximal canonical representation for the spectral covariance distribution $\underline{\underline{B}}$ of $\underline{\underline{x}}_n$, $n \in Z$, in the form (9a-c). Then for any integer $p > 0$ the best p-step linear least square predictions

$$\hat{x}_{n,(p)}^{(j)} = P_{n-p}\, x_n^{(j)}, \qquad j = 1,\dots,q; \quad n \in Z,$$

form such a sequence of random vectors $\hat{\underline{\underline{x}}}_{n,(p)}$, $n \in Z$, that

(i) $\hat{\underline{\underline{x}}}_{n,(p)}$, $n \in Z$, satisfies (3),

(ii) a representation for the spectral covariance distribution $\hat{\underline{\underline{B}}}_{(p)}$ of $\hat{\underline{\underline{x}}}_{n,(p)}$, $n \in Z$, is

$$\hat{\underline{\underline{B}}}_{(p)} = \sum_{k=-\infty}^{\infty} \hat{\underline{\underline{U}}}_{k,(p)} \otimes \hat{\underline{\underline{U}}}_{k,(p)}^* ,$$

where the components of $\hat{\underline{\underline{U}}}_{n,(p)} = (\hat{U}_{n,(p)}^{(hj)})_{q \times r_n}$, $n \in Z$, are defined according to (18).

Proof. Since $E|x_{n,(p)}^{(j)}|^2 \leq E|x_n^{(j)}|^2$ for all $j = 1,\dots,q$; $n \in Z$, it is clear that the sequence $\hat{\underline{\underline{x}}}_{n,(p)}$, $n \in Z$, satisfies (3).

The proof of (ii) is based on Theorem 7.

Since the representation (19) for $\underline{\underline{B}}$ is maximal canonical, it follows from Theorem 7 that there exists a Cramér-Hida representation

$$(20) \qquad x_n^{(j)} = \sum_{k \in I_n} \sum_{s=1}^{r_k} U_k^{(js)}(\chi_n)\, \tilde{\eta}_k^{(s)} , \quad j = 1,\dots,q; \quad n \in Z,$$

for $\underline{\underline{x}}_n$, $n \in Z$, in the form (2); here $I_n = \{ k \mid m(k) \leq n \}$. By applying Theorem 8 to the representation (20), it then follows that for any $n \in Z$, $j = 1,\ldots,q$

$$\hat{\underline{\underline{x}}}^{(j)}_{n,(p)} = \sum_{k \in I_{n,(p)}} \sum_{s=1}^{r_k} U_k^{(js)}(\chi_n)\, \tilde{\eta}_k^{(s)} ,$$

where $I_{n,(p)} = \{k \mid m(k) \leq n - p\}$. Furtheremore, as in Example 9, we get

$$\hat{x}^{(j)}_{n,(p)} = \sum_{k \in I_{n,(p)}} \sum_{s=1}^{r_k} \hat{U}^{(js)}_{k,(p)}\, (\chi_n)\, \tilde{\eta}_k^{(s)} ,$$

where $\hat{U}^{(js)}_{m,(p)}$, $s = 1,\ldots,r_m$; $m \in Z$, are defined according to (18). Thus

$$\hat{\underline{\underline{B}}}_{(p)} = \sum_{k=-\infty}^{\infty} \hat{\underline{\underline{U}}}_{k,(p)} \otimes \hat{\underline{\underline{U}}}^*_{k,(p)} \tag{21}$$

is a representation for the covariance distribution $\hat{\underline{\underline{B}}}_{(p)}$ of $\hat{\underline{\underline{x}}}_{n,(p)}$, $n \in Z$.

The theorem is proved.

<u>Remark.</u> The exist such purely non-deterministic sequences of random vectors $\underline{\underline{x}}_n$, $n \in Z$, satisfying the condition (3), that the representation (21) for $\hat{\underline{\underline{B}}}_{(p)}$, corresponding to a maximal canonical representation (19) for $\underline{\underline{B}}$, is not canonical.

<u>Remark</u>. Let $\underline{\underline{x}}_n$, $n \in Z$, be a purely non-deterministic sequence of random vectors satisfying (3) and let

$$\underline{\underline{B}} = \sum_{k=-\infty}^{\infty} \underline{\underline{U}}_k \otimes \underline{\underline{U}}^*_k$$

be a maximal canonical representation for its spectral covariance distribution $\underline{\underline{B}}$ in the form (9a–c). In then follows from Theorem 8 that the p-step prediction error is

$$\sigma_{n,(p)}^{(j)} = \sum_{k \in I'_{n,(p)}} \sum_{s=1}^{r_k} | U_k^{(js)}(\chi_n) |^2 ,$$

where $I'_{n,(p)} = \{ k \mid n-p < m(k) \leq n \}$.

5. The stationary case

In this section we present a relationship between a maximal canonical representation for the spectral covariance distribution $\underline{\underline{B}}$ of a purely non-deterministic stationary sequence of random vectors $\underline{\underline{x}}_n$, $n \in Z$, and a maximal matrix related to the spectral density matrix of $\underline{\underline{x}}_n$, $n \in Z$.

The terminology concerning purely non-deterministic (or linearly regular) stationary sequence of random vectors is adapted from Rozanov [6].

First we consider the representation (9a-c) for the spectral covariance distribution in the stationary case.

Theorem 11. Let $\underline{\underline{x}}_n$, $n \in Z$, be a sequence of random vectors satisfying the growth condition (3). Then $\underline{\underline{x}}_n$, $n \in Z$, is a purely non-deterministic stationary sequence and the rank of $\underline{\underline{x}}_n$, $n \in Z$, is r if and only if there exists a canonical representation for the spectral covariance distribution $\underline{\underline{B}}$ of $\underline{\underline{x}}_n$, $n \in Z$, in the form

$$\underline{\underline{B}} = \sum_{k=-\infty}^{\infty} (\overline{\chi}_k \underline{\underline{\Phi}}) \otimes (\overline{\chi}_k \underline{\underline{\Phi}})^* , \tag{22}$$

where $\underline{\underline{\Phi}}(\lambda) = (\Phi^{(hj)}(\lambda))_{q \times r}$, $\lambda \in [0,1]$, is a matrix of square integrable functions on $[0,1]$ such that for all $h = 1,\dots,q$; $j = 1,\dots,r$,

$$\int_0^1 \chi_n(\lambda)\, \Phi^{(hj)}(\lambda)\, d\lambda = 0 \quad \text{when} \quad n < 0.$$

If (22) is a canonical representation for $\underline{\underline{B}}$, then

$$\underline{\underline{f}}(\lambda) = \underline{\underline{\Phi}}(\lambda)\, \underline{\underline{\Phi}}(\lambda)^* , \qquad \lambda \in [0,1] ,$$

is the spectral density matrix of $\underline{\underline{x}}_n$, $n \in Z$; and the rank of

$\underline{\underline{f}}(\lambda)$ is r for almost all $\lambda \in [0,1]$.

Proof. Suppose $\underline{\underline{x}}_n$, $n \in Z$, is a purely non-deterministic stationary sequence of random vectors and suppose the rank of $\underline{\underline{x}}_n$, $n \in Z$, is r.

The Ceamér-Hida representation (2) for $\underline{\underline{x}}_n$, $n \in Z$, has then the form

$$x_n^{(j)} = \sum_{k=-\infty}^{n} \sum_{s=1}^{r} c_{n-k}^{(sj)} \eta_k^{(s)}, \quad j = 1,\ldots,q; \quad n \in Z$$

(Rozanov [6; pp.41-42]). Thus, it follows from Theorem 7 that

$$\underline{\underline{B}} = \sum_{k=-\infty}^{\infty} (\overline{\chi}_k \underline{\underline{\Phi}}) \otimes (\overline{\chi}_k \underline{\underline{\Phi}})^* ,$$

is a (maximal) canonical representation for the covariance distribution $\underline{\underline{B}}$ of $\underline{\underline{x}}_n$, $n \in Z$; here

$$\underline{\underline{\Phi}} = (\Phi^{(hj)})_{q \times r}$$

is a matrix of square integrable functions $\Phi^{(hj)}: [0,1] \longrightarrow C$ for which

$$\int_0^1 \chi_n(\lambda)\, \Phi^{(hj)}(\lambda)\, d\lambda = c_n^{(hj)}$$

for all $h = 1,\ldots,q$; $j = 1,\ldots,r$; $n \in Z$. (Note that

$$\sum_{k=0}^{\infty} \sum_{s=1}^{r} |c_k^{(js)}|^2 = E\,|x_0^{(j)}|^2 < \infty$$

for all $j = 1,\ldots,q$.)

On the other hand, suppose (22) is a canonical representation for the spectral covariance distribution $\underline{\underline{B}}$ of $\underline{\underline{x}}_n$, $n \in Z$.

It is clear that the sequence $\underline{\underline{x}}_n$, $n \in Z$, is stationary and purely non-deterministic. Moreover, since the representation (22) is canonical, the rank of $\underline{\underline{x}}_n$, $n \in Z$, is r.

Furthermore, a straightforward application of Lemma 2 shows that the matrix

$$\underline{\underline{f}}(\lambda) = \underline{\underline{\Phi}}(\lambda) \ \underline{\underline{\Phi}}(\lambda)^* , \quad \lambda \in [0,1],$$

is the spectral density matrix of $\underline{\underline{x}}_n$, $n \in Z$; and a fortiori the rank of $\underline{\underline{f}}(\lambda)$ is r for almost all $\lambda \in [0,1]$ (cf. Rozanov [6; p. 57]).

The theorem is proved.

Remark. Let $\underline{\underline{x}}_n$, $n \in Z$, be a purely non-deterministic stationary sequence of random vectors of rank r and let $\underline{\underline{f}}(\lambda)$, $\lambda \in [0,1]$, be the spectral density matrix of $\underline{\underline{x}}_n$, $n \in Z$. Then to any representation

$$\underline{\underline{f}}(\lambda) = \underline{\underline{\Phi}}(\lambda) \ \underline{\underline{\Phi}}(\lambda)^*, \quad \lambda \in [0,1],$$

where $\underline{\underline{\Phi}}(\lambda) = (\Phi^{(hj)}(\lambda))_{q \times r}$, are rectangular matrices for (almost) all $\lambda \in [0,1]$, it corresponds a canonical representation for the spectral covariance distribution $\underline{\underline{B}}$ of $\underline{\underline{x}}_n$, $n \in Z$, in the form (22) (cf. Rozanov [6; Chapter I, Theorem 9.1]).

The following theorem follows by combining the results stated in Theorem 7, Theorem 11 and Theorem 4.3 in Rozanov [6; pp.60-61].

Theorem 12. Let $\underline{\underline{x}}_n$, $n \in Z$, be a stationary purely non-deterministic sequence of random vectors of rank r. Then a canonical representation

$$\underline{\underline{B}} = \sum_{k=-\infty}^{\infty} (\overline{X}_k \ \underline{\underline{\Phi}}) \otimes (\overline{X}_k \ \underline{\underline{\Phi}})^* \tag{23}$$

for the spectral covariance distribution $\underline{\underline{B}}$ of $\underline{\underline{x}}_n$, $n \in Z$, where $\underline{\underline{\Phi}}(\lambda) = (\Phi^{(hj)}(\lambda))_{q \times r}$, $\lambda \in [0,1]$,

$$\Phi^{(hj)} = \sum_{k=0}^{\infty} c_k^{(hj)} \ \overline{X}_k$$

($c_0^{(hj)} \neq 0$ for some pair (hj)), is maximal, if and only if the matrix $\underline{\underline{\Phi}}(0) \ \underline{\underline{\Phi}}(0)^*$ is maximal in the sense of Rozanov [6; p.60]; here $\underline{\underline{\Phi}} = (\Phi^{(hj)})_{q \times r}$,

$$\Phi^{(hj)}(z) = \sum_{k=0}^{\infty} c_k^{(hj)} \ z^k, \quad |z| < 1.$$

Remark. Let $\underline{\underline{x}}_n$, $n \in Z$, be a stationary purely non-deterministic sequence of rank r. Suppose (23) is maximal canonical representation for the spectral covariance distribution $\underline{\underline{B}}$ of $\underline{\underline{x}}_n$, $n \in Z$. Then

$$(24) \qquad \hat{\underline{\underline{B}}}_{(p)} = \sum_{k=-\infty}^{\infty} (\overline{\chi}_k \hat{\underline{\underline{\Phi}}}_{(p)}) \otimes (\overline{\chi}_k \hat{\underline{\underline{\Phi}}}_{(p)})^* ,$$

where the elements of the matrix $\hat{\underline{\underline{\Phi}}}_{(p)}$ are defined according to (18), is a representation for the spectral covariance distribution $\hat{\underline{\underline{B}}}_{(p)}$, of the best p-step linear least square predictions $\underline{\underline{x}}_{n,(p)}$, n Z, of $\underline{\underline{x}}_n$, n Z. There exist such stationary purely non-deterministic sequences $\underline{\underline{x}}_n$, n Z, that the representation (24) for $\underline{\underline{B}}_{(p)}$, corresponding to a maximal canonical representation (23) for $\underline{\underline{B}}$, is not canonical.

References

[1] Abreu, J.L., H-valued generalized functions and orthogonally scattered measures. Advances in Math. 19 (1976), 382-412

[2] Cramér, H., On some classes of non-stationary stochastic processes. In Proceedings of the Fourth Berkeley symposium on mathematical statistics and probability. II, pp.57-58. University of California Press, Berkeley 1962.

[3] Deo, D.M, Prediction theory of non-stationary processes. Sankhya Ser. A 27 (1965), 113-132.

[4] Dudley, R.M., Prediction theory for non-stationary sequences. Proceedings of the Fifth Berkeley symposium on mathematical statistics and probability. II: 1, pp.223-234. University of California Press, Berkeley, 1967.

[5] Masani, P., The prediction theory of multivariate stochastic processes, III. Acta Math. 104 (1960), 141-162.

[6] Rozanov, Yu. A., Stationary random processes. English translation. Holden-Day series in time series analysis. Holden-Day, San Francisco, 1967.

[7] Schwartz, L., Théorie des distribution. Nouvelle édition. Hermann, Paris, 1966.

[8] Wiener, N., and P. Masani,. The prediction theory of multivariate stochastic processes, Part I. Acta Math. 98 (1957), 111-150; Part II. Acta Math. 99 (1958), 93-137.

University of Helsinki
Department of Mathematics
SF-00100 Helsinki 10

F i n l a n d

AN INEQUALITY FOR THE SEMI-MARTINGALES

APPLICATION TO STOCHASTIC DIFFERENTIAL EQUATIONS

J. Pellaumail

Summary :

We prove the existence and unicity of a strong solution for the stochastic differential equation $dX_t = a(X,t)dt$, where a is "predictable", locally bounded and locally lipschitzian and X is a semi-martingale (cf. [3]). The proof uses an inequality for such a semi-martingale which is established in the paragraph B. There results have been shown in [2] and generalize [1].

A - STOCHASTIC DIFFERENTIAL EQUATIONS

A-1. Generalities:

In this paragraph A, we consider: $T = [0,1]$, the unit interval of the real line

- H and K two separable Banach spaces and $\mathcal{L}_a(K,H)$ a subspace of $\mathcal{L}(K,H)$, the space of the linear operators from K to H; on this subspace $\mathcal{L}_a(K,H)$, we consider a norm such that, if u is an element of $\mathcal{L}_a(K,H)$, then $||u|| \geq \sup_{||k||\leq 1} ||u(k)||_H$

- $(\Omega,\mathcal{F},P,(\mathcal{F}_t)_{t \in T}) = B^I$ a "stochastic basis" with the usual assumption, i.e. for each element t of T, $\mathcal{F}_t = \bigcap_{s>t} \mathcal{F}_s$ and $A \in \mathcal{F}_t$ if $P(A) = 0$. We shall call this basis the "initial basis". Denote by $\mathcal{A}$ the algebra generated by the sets $F \times]s,t]$ with $F \in \mathcal{F}_s$; the σ-algebra generated by $\mathcal{A}$ is the σ-algebra of predictable sets.

A-2. Canonical basis (Definition)

We shall use the french notations "cadlag" "caglad", and so on ; more precisely, let f be a real function defined on T; we say that f is cadlag if, for each element of T,f is right continuous and has left limit. (in french : continu à droite et a une limite à gauche). We say that a process X is cadlag if, for each element ω of Ω , the sample function $t \longrightarrow X_t(\omega)$ is cadlag.

Let D^H be the space of all H-valued cadlag functions defined on T. For each element t of T, let $\mathcal{D}_t^H$ be the σalgebra generated by the sets $\{\omega: X_s(\omega) \in H_0\}$ with $s \leq t$ and H_0 borelian set of H; we define $\mathcal{D}^H = \mathcal{D}_1^H$. The family

$(D^H \times \Omega, \mathcal{D}^H \otimes \mathcal{F}, (\mathcal{D}_t^H \otimes \mathcal{F}_t)_{t \in T})$ is a "basis of process" that we shall note B^H and which we shall call the canonical basis (for the H-valued processes defined with respect to the basis B^I).

A-3. Remarks and conventions :

a) The σ-algebra od predictable sets of the canonic basis B^I is generated by the sets $G \times F \times]s,t]$ where G is an element of $\mathcal{D}_s^H$, and F is an element of $\mathcal{F}_s$; actually, it is sufficient to consider the case where G is the set of the cadlag functions x such that $x_u = x(u)$ is an element of H_0 with $u < s$ and H_0 borelian set of H.

b) Let $a(x,\omega,t)$ be an $\mathcal{L}_a(H,H)$-valued process defined with respect to the canonic basis B^H. Let X be an H-valued process defined with respect to the initial basis B^I. In the sequel we consider processes Z such that $Z_t(\omega) = a[X(\omega),\omega,t]$; in this situation, for the commodity of notations, we shall not write the symbol ω; then, we shall write $Z_t = a(X,t)$.

c) we shall consider stopping times such that :

$$w = \inf\{t: \ t \geq u, \ t \leq v, \ \|X_t\| > \varepsilon\}$$

In this situation, if the set above is empty, we define $w(\omega) = v(\omega)$.

d) Let u and v be two stopping times. We define the stochastic integral $\int_{]u,v]} Y \cdot dX$ as usual and we define :

$$\int_{]u,v[} Y \cdot dX = \int_{]u,v]} Y \cdot dX - Y_{v-}(X_v - X_{v-})$$

(when these terms are well defined).

If v is a predictable stopping time, the set $]u,v[$ is a predictable set, then we have :

$$\int_{]u,v[} Y \cdot dX = \int 1_{]u,v[} \, Y \cdot dX$$

e) Let u be a stopping time and let X and Y be two processes, the process X being cadlag ; then we shall denote by

$\operatorname{Sup}_{t<v} \| \int_{]o,t]} Y_s \cdot dX_s \|^2$ the random variable U defined more precisely by :

$$U(\omega) = \operatorname{Sup}_{t<u(\omega)} \| Z_t(\omega) \|^2$$

where Z is defined by

$$Z_t = \int_{]o,t]} Y_s \cdot dX_s$$

A-4. <u>Proposition.</u>

Let K be a Banach space. Let X be a cadlag H-valued process, defined and adapted with respect to the initial basis B^I. Let $a(x,\omega,t)$ be a K-valued process, defined and predictable with respect to the canonical basis B^H. Let Y be the process defined by : $Y_t(\omega) = a(X(\omega),\omega,t)$. Then, Y is a K-valued process, predictable with respect to the initial basis B^I. Moreover, $Y_t(\omega)$ is depending only on the values $X_s(\omega)$ for $s < t$ (then, it is possible to define $Y_t(\omega)$, when X_s is known only for $s < t$).

<u>Proof.</u> 1°) First, we consider the case where there exists k element of K, $u < v < w$ elements of T, H_o-borelian subset of H, F-element of $\mathcal{F}_v$ such that, if

$$J = \{x: \ x_u \in H_o\}$$

then, $a(x,\omega,t) = k \cdot 1_J(x) \cdot 1_F(\omega) \cdot 1_{]v,w]}(t)$

Let F' be the set defined by $F'=\{\omega : X_u(\omega) \in H_o\}$. The process X being adapted F' belongs to $\mathcal{F}_u$; we have also :

$$Y_t(\omega)=a[X.(\omega),\omega,t] = k1_J[X.(\omega)] \cdot 1_F(\omega) \cdot 1_{]v,w]}(t)$$

$$= k \cdot 1_F(\omega) \cdot 1_{F'}(\omega) \cdot 1_{]v,w]}(t)$$

then Y is a predictable process and $Y_t(\omega)$ is depending only on $X_s(\omega)$ for $s < t$.

2°) Then, we consider an H-valued process X, adapted with respect to the initial basis B^I. Let $\mathcal{C}_X$ be the family of all the K-valued processes a, defined with respect to the canonical basis B^H and such that, if $Y = a(X,t)$, Y is a predictable process with Y_t depending only on X_s for $s < t$. The space $\mathcal{C}_X$ is a vector space and a monotone class ; moreover, $\mathcal{C}_X$ contains all the processes $a=k \cdot 1_J \cdot 1_F \cdot 1_{]v,w]}$ as defined in the 1°) above.

Then, $\mathcal{C}_X$ contains all the predictable processes (cf. the remark A-3-a).

A-5 : **Theorem.** Let H and K be two separable Hilbert spaces. Let $B^I = (\Omega, \mathcal{F}, P, (\mathcal{F}_t)_{t \in T})$ be a "stochastic basis", with the usual assumptions, (cf. A-1-above), that we shall call the initial basis. Let Z be a K-valued cadlag process, defined and adapted with respect to the initial basis B^I. We suppose that there exists a real positive increasing process Q, defined and adapted with respect to the initial basis B^I, such that, for each (strongly) predictable $\mathcal{L}_a(K,H)$-valued uniformly bounded process Y, and for each stopping time u, we have (cf. A-3-e above) :

$$\text{(i)} \quad E(\operatorname*{Sup}_{t<u}||\int_{]o,t]} Y_s \cdot dZ_s||^2) \leq E(Q_{u-} \cdot (\int_{]o,u[} ||Y_t||^2 \cdot dQ_t))$$

Let $a(x,\omega,t)$ be an $\mathcal{L}_a(K,H)$-valued process, defined and predictable with respect to the canonical basis B^H. We suppose that a is locally lipschitzian and locally bounded in the following sense :

(ii) For each real positive number d, there exists a real number L_d such that, if (ω,t) is an element of $(\Omega \times T)$, if (x,x') is a pair of elements of D^H with $\operatorname*{Sup}_{s\leq t}||x_s|| \leq d$ and $\operatorname*{Sup}_{s\leq t}||x'_s|| \leq d$, then we have :

$$||a(x,\omega,t)-a(x',\omega,t)||^2 \leq L_d \cdot \operatorname*{Sup}_{s<t}||x_s-x'_s||^2$$

(iii) For each real positive number d, there exists a real number C_d such that, if (x,ω,t) is an element of $(D^H \times \Omega \times T)$, with $\operatorname*{Sup}_{s<t}||x_s|| \leq d$, then we have $||a(x,\omega,t)|| \leq C_d$.

Let u be a stopping time and let X_u be an H-valued and $\mathcal{F}_u$-measurable random variable.
Then, there exists a predictable stopping time v and an H-valued cadlag process X, defined and adapted with respect to the initial basis B^I, unique up to indistinguishability, with the following properties :

(iv) if ω belongs to the set $\{v < 1\}$, then

$$\operatorname*{lim.sup.}_{t\uparrow v(\omega)} ||X_t(\omega)|| = +\infty$$

$$\text{(v)} \quad X_t = X_u + \int_u^t a(X,s) \cdot dZ_s$$

on the stochastic interval $]u,v[$, this integral being an usual stochastic integral.

Then, we say that X is a strong solution of the differential equation $dX_t = a(X,t).\, dZ_t$ on a the stochastic interval $]u,v[$, with the initial value X_u.

Proof. By localization, it is sufficient to consider the case where Q is uniformly bounded; then, in the right term of the inequality A-5-(i), we can write:

$$E\{\int_{]0,u[} ||Y_t||^2 \,.\, dQ_t\} \quad \text{instead of } E(Q_{u-} \,.\{\int_{]0,u[} ||Y_t||^2 \,.\, dQ_t\})$$

that we shall do henceforth. In the sequel we shall omit the symbol ω if there is no possible confusion. The following proof is a natural generalization of the clasical study of ordinary differential equations based on fixed point theorem. This proof has three steps :

1°) Unictly : Lemma A-6

2°) Existence of a local solution (A-7)

3°) Building of the "global solution". (A-8).

A-6 Unicity. We consider the hypothesis and notations given in the theorem A-5 above. Let X and X′ be two adapted cadlag processes which are solutions of the equation A-5-(v) on the stochastic intervals $]u,v]$ and $]u,v']$ respectively and which have the same initial value $X_u = X'_u$

Then, $X.1_{[u,v\ v']}$ and $X'.\ 1_{[u,v\ v']}$ are two indistinguishable processes.

Proof. Let X and X′ be two solutions on $]u,v]$ and $u' = \inf\{t: ||X_t - X'_t|| > 0,\ t \geq u,\ t \leq (v \wedge v')\}$.
If $P[u' < (v\wedge v')] = 0$, the lemma is proved. Then, we suppose that $P[u' < (v\wedge v')] > 0$. The processes X and X′ being cadlag, there are a real number d and a stopping time w′ such that:

$$\sup_{u' \leq s \leq w'} (||X_s|| + ||X'_s||) \leq d$$

$$P([w' > u') > 0 \text{ and } w' \leq (v\ v').$$

Let L_d be the "Lipschitz real number", associated to d, which appears in the condition A-5-(ii) Let w be stopping time defined by :

$$w = \inf.\{t : t \geq u' , t \leq w', Q_t - Q_{u'} > \frac{1}{2L_d}\}$$

The process Q being right continuous, we have $P([w > u']) > 0$. Then we define :

$$h = E\{\sup_{u \leq s \leq w} ||X'_s - X_s||^2\}$$

Then, we have (cf. A-3-e) :

$$h = E\{\sup_{u \leq s \leq w} || \int_{]u,s]} [a(X,r)-a(X',r)] \cdot dZ_r||^2\}$$

$$\leq E\{ \int_{]u,w[} ||a(X,r)-a(X',r)||^2 \cdot dQ_r\} \quad (\text{cf.A-5-(i)})$$

$$\leq E\{\int_{]u,w[} L_d \cdot \sup_{u \leq r < w} \cdot ||X_r - X'_r||^2 \cdot dQ_r\} \quad (\text{cf.A-5-(ii)})$$

$h \leq \frac{1}{2} h$ (cf. the building of w); then $h = 0$ and that proves the unicity.

A-7 __Local Solution.__ We consider the hypothesis and notations given in the theorem A-5, Then, there exist a stopping time v and an H-valued cadlag adapted process X, defined on the stochastic interval $[u,v]$, which satisfy the following properties :

(i) $P(\{v > u\}) > 0$

(ii) $X_t = X_u + \int_u^t a(X,s)dZ_s$ on the stochastic interval $\{(t,\omega) : u(\omega) \leq t \leq v(\omega)\}$

__Proof.__ 1°) Let X^0 be the process defined by :

$$X^o_t(\omega) = 1_{[u,1]}(t,\omega) \cdot X_u(\omega)$$

Let d be a real number such that

$$P[||X_u|| > d] < 1$$

Let v'_0 be the stopping time defined by :

$$v'_0(\omega) = \inf. \{t : ||X^o_t|| > d \}$$

$$(v'_0(\omega) = u(\omega) \quad \text{or} \quad v'_0(\omega) = 1)$$

Let $L = L_{2d}$ be the "Lipschitz real number" associated to 2d which appears in A-5-(ii).

Let $v''(0)$ be the stopping time defined by :

$$v''(0) = \inf.\{t:\ t \geq u,\ \ t \leq v'_0,\ Q_t - Q_u > \frac{1}{8L}\}$$

The processes Q being right continuous, we have $P([v''(0) > u]) > 0$.

Let X^1 be the cadlag process defined on the stochastic interval $[u,v''(0)]$ by :

$$X^1_t = X^0_t + \int_{]u,t]} a(X^0,s)\ .\ dZ_s$$

Let v(0) we the stopping time defined by :

$$v(0) = \inf.\{\ t:\ t \geq u,\ \ t \leq v''(0),\ ||X^1_t - X^0_t|| > \frac{d}{4}\ \}$$

The processes X^1 and X^0 being right continuous, $P([v(0) > u]) > 0$: we have also :

$$P([v(0) > u]) \geq \frac{16}{d^2}\ .\ E\ \{\ \sup_{u \leq s < v(0)} ||X^1_s - X^0_s||^2\ \}$$

Then, we define, by recurrence, the sequence $(v(n))_{n>0}$ of stopping times and the associated sequence $(X^n)_{n>0}$ of H-valued cadlag processes. in the following way :

$$v(n) = \inf.\{t\ :\ t \geq u,\ t \leq v(n-1),||X^n_t - X^{n-1}_t|| > 2^{-n}.d\}$$

On the stochastic interval $]u,v(n)[$, the process $||X^n||$ is bounded by 2d, then, we can define, on the stochastic interval $]u,v(n)]$ the process X^{n+1} by :

$$X^{n+1}_t = X^0_t + \int_{]u,t]} a(X^n,s)\ .\ dZ_s$$

We define the stopping time v by:

$$v = \lim_{n\to\infty}.v(n)$$

(the sequence $(v(n))_{n>0}$ is decreasing to v). Now, we shall prove (3°) below that $P([v > u]) > 0$

2° We first prove an inequality. We define :

$$h_n = E\ \{\ \sup_{u \leq s < v(n)} ||X^{n+1}_s - X^n_s||^2\ \}\ .\ \text{We have}$$

$$h_n = E\ \{\sup_{u \leq s < v(n)} ||\int_{]u,s]} [a(X^n,r)].\ dZ_r||^2\ \}$$

(cf.A-3-e)

$$\leq E \{ \int_{]u,v(n)[} ||a(X^n,r)-a(X^{n-1},r||^2 \cdot dQ_r$$

(cf.A-5-(i))

$$\leq E \{ \int_{]u,v(n)[} L_{2d} \cdot \sup_{u\leq r<v(n)} ||X_r^n - X_r^{n-1}||^2 \cdot dQ_r \}$$

(cf.A-5-(ii)

$$h_n \leq L \cdot h_{n-1} \cdot \frac{1}{8L} \leq \frac{1}{8} h_{n-1} \leq 8^{-n} h_o$$

$$h_n \leq 8^{-n} \cdot \frac{d^2}{16} \cdot P([v(0) > u]).$$

3° Let $A(n)$ be the set defined by :

$$A(n) = \{\omega : [v(n)](\omega) < [v(n-1)](\omega) \}$$

If $\omega \in A(n)$, we have :

$$||X^n_{[v(n)](\omega)} - X^{n-1}_{[v(n)](\omega)}||(\omega) > 2^{-n} \cdot d, \text{ then}$$

$$h_{n-1} = E \{ \sup_{u\leq s<v(n-1)} ||X_s^n - X_s^{n-1}||^2\} \geq 4^{-n} \cdot d^2 \cdot P[A(n)]$$

$$P[A(n)] \leq 8 \cdot 8^{-n} \cdot h_o \cdot 4^n \cdot \frac{1}{d^2} \leq 2^{-(n+1)} \cdot P([v(0) > u])$$

But $\quad [v < v(0)] \subset \bigcup_{n\geq 1} [v(n) < v(n-1)]$, then

$$P[v < v(0)] \leq \sum_{n\geq 1} P[A(n)] \leq \frac{1}{2} \cdot P([v(0) > u])$$

$$P([v > u]) \geq \frac{1}{2} P([v(0) > u]) > 0$$

4° Now, we can build the process X, On the stochastic interval $]u,v[$, $||X^n - X^{n-1}||$ is bounded by $2^{-n}d$; then, on this set, the sequence $(X^n)_{n>0}$ converges uniformly to an adapted cadlag process X uniformly bounded by 2d. Then, on the stochastic interval $]u,v]$, $||a(X^n,t)-a(X^{n-1},t)||$ is bounded by $L.2^{-n}.d$; then, on this set, the sequence $(a(X^n,t))_{n>0}$ of predictable processes converges uniformly to the predictable process $a(X,t)$

Moreover, for each n, on the stochastic interval $]u,v]$, we have :

$$X_t^{n+1} = X_t^o + \int_{]u,t]} a(X^n,s) \cdot dZ_s$$

By passing to the limit, we obtain :

$$X_t = X_t^o + \int_{]u,t]} a(X,s) \cdot dZ_s$$

on the stochastic interval $]u,v]$.

A-8. <u>Global solution</u>. Now, we shall prove the theorem A-5. We consider the hypothesis given in A-5.

1°) Let m be the supremum of the real numbers m_n such that there exist a stopping time $u(n)$ and a process X^n with :

$m_n = (P \otimes \mu)\ (u,u(n)])$ where μ is the Lebesque measure and $X_t^n = X_u + \int_u^t a(X^n,s).dZ_s$ on the stochastic interval $]u,u(n)]$.

Let $(m_n,u(n),X^n)_{n>0}$ be a sequence of triples of real numbers, stopping times and processes associated as above, such that $m_n \uparrow m$. The processes X^n and X^{n+1} are indistinguishable on the stochastic interval $[u,u(n) \wedge u(n+1)]$ (cf.A-6) ;

Then, if $v'(n) = \operatorname{Sup.}_{k\leq n} u(k)$, we can define the process Y^n on $[u,v'(n)]$ by $Y^n = X^k$ on $[u,u(k)]$ for each k such that $k \leq n$.

Let v' be the stopping time defined by $v' = \operatorname{Sup}_{n>0} v'(n)$. Let X on $[u,v'[$ by $X = Y^n$ on $[u,v'(n)]$. Let d be a real positive number. For each integer n, let $v(n)$ be the stopping time defined by :

$$v(n) = \inf.\{t : t \leq v'(n) \text{ and } ||Y_t^n|| > \quad \}$$

We define $v = \operatorname{Sup.}_v v(n)$ and we first shall prove that

$P(\{v < v' \text{ or } v' = 1\}) = 1$. We suppose that there exists a real number d such that:

$$P(\{v = v' \text{ and } v' < 1\}) > 0$$

and we shall prove that there is an impossibility.

2°) The process X is defined and bounded in norm by d on the stochastic interval $]u,v[$ (cf. the definition of v); then, the process $a(X,t)$ is bounded in norm by C_d and we can define:

$$\hat{X}_t = X_u + \int_{]u,t]} a(X,s) \cdot dZ_s \quad \text{on } [u,v]$$

The processes $\hat{X}$ and X are indistinquishable on the interval

$$[u,v] \cap \{ \bigcup_{n>0} [u,v'(n)]\} \qquad \text{(cf. A-6) ;}$$

then, it is possible to define X on the interval $[u,v]$, by $X = \hat{X}$ and X is well defined on the interval $[u,v] \cup \{\bigcup_{n>0}[u,v'(n)]\}$.
On this predictable interval, X is a solution of A-5.(v)
Then, we define :

$$w(\omega)=\begin{cases} v(\omega) & \text{if} \quad v(\omega) = v'(\omega) \\ 1 & \text{if} \quad v(\omega) < v'(\omega) \end{cases}$$

$$w'(\omega)=\begin{cases} v'(\omega) & \text{if} \quad v(\omega) < v'(\omega) \\ 1 & \text{if} \quad v(\omega) = v'(\omega) \end{cases}$$

and $\bar{a} = a - a.\, 1_{]v,w]}$

By the lemma A-7, it is possible to build a stopping time w'' with $P([w'' > w]) > 0$ and an associated process $\bar{X}$, solution of

$$\bar{X}_t = X_u + \int_{]o,t]} \bar{a}(\bar{X},s) \cdot dZ_s \quad \text{on} \quad [u,w'']$$

Then, we can extend the process X on $[u,w''] \cap [u,w'[$ by defining $X = \bar{X}$ on $[u,w'' \wedge w']$ (cf.A-6); but, this is impossible by the definition of m.
Then, we have necessarily $P(\{v < v'\}) = 1$ (for each real number d).

3°) Let $(w(n))_{n>0}$ be the sequence of stopping times defined by :

$$w(n) = \text{inf.} \{t: \quad t \geq u, \quad t \leq v, \; \|X_t\| > n, \; t \leq 1 - \tfrac{1}{n}\}$$

The sequence is increasing to v and for each integer n, $P([w(n) < v]) = 1$; then v is a predictable stopping time and that completes the proof.

A-9. <u>Remark.</u> Let c be a measurable mapping from $(H \times T)$ into a Banach space K which is continuous with respect to the first variable. For each element (x,ω,t) of $(D^H \times \Omega \times T)$, we put $a(x,\omega,t) = \lim. c(x_s,t)$ (actually, a does not depend on ω) ; it is easily seen that a is well defined and is K-valued predictable process with respect to the canonical basis B^H. Thus, this situation is a particular of the situation studied before.

B - AN INEQUALITY FOR SEMI-MARTINGALES

B-1. Generalities. The main result of this paragraph is the theorem B-6 below. It is fundamental to note that this inequality B-6-(i) is only concerned by the values of processes Z and A "strictly before" the stopping time u. The theorem B-6 gives an example where we have the condition (i) of the theorem A-5 above.

In this paragraph, we shall consider the hypothesis and notations given in A-1 and we shall use the french notation cadlag and the conventions given in A-3-d) and A-3-e). Moreover, if M is an H-valued square integrable cadlag martingale, we denote by $[M]$ the increasing positive cadlag adapted process which is the quadratic variation of M, i.e.

$$[M]_t = \lim_{n\to\infty}\left(\sum_{k=0}^{\infty} ||M_{(k+1).2^{-n}\wedge t} - M_{k.2^{-n}\wedge t}||^2 \right)$$

and we shall denote by $<M>$ the "natural" process associated to $[M]$ (i.e. the predictable increasing process such that $[M]_0 = 0$ and $[M]_t - <M>_t$ is a real martingale).

B-2. A lemma on the conditional expectations. We consider $(\Omega, \mathcal{F}, P)$ a probability space, $\mathcal{G}$ a sub-σ-algebra of $\mathcal{F}$, A an element of $\mathcal{F}$ and $\mathcal{G}^*$ the σ-algebra generated by $\mathcal{G}$ and A. Let Z be an element of $L_1^H(\Omega, \mathcal{G}^*, P)$ such that $E(Z|\mathcal{G}) = 0$.
If we write $B = \Omega \setminus A$, we have :

(i) $E(1_A|\mathcal{G}).E(||Z||^2 . 1_A| \ ) = E(1_B|\mathcal{G}).E(||Z||^2 . 1_B|\mathcal{G})$ a.e.

(ii) $E(1_B||Z||^2) = E\{1_A . E(||Z||^2 | \mathcal{G})\}$

Proof. 1° The following elementary proof was suggested by J. Jacod.
We can write $Z = X . 1_A + Y . 1_B$ where X and Y belong to $L_1^H(\Omega, \mathcal{G}, P)$. The property $E(Z|\mathcal{G}) = 0$ implies

$$X . E(1_A|\mathcal{G}) = E(X . 1_A|\mathcal{G}) = E(Y . 1_B|\mathcal{G}) = -Y . E(1_B|\mathcal{G}).$$

Then, we have also :

$$E(1_A|\mathcal{G}) \cdot E(||Z||^2 \cdot 1_A|\mathcal{G}) = E(1_A|\mathcal{G}) \cdot E(||X||^2 \cdot 1_A|\mathcal{G})$$
$$= [E(1_A|\mathcal{G})]^2 \ ||X||^2 = ||Y||^2 \cdot [E(1_B|\mathcal{G})]^2$$
$$= E(1_B|\mathcal{G}) \cdot E(||Z||^2 \cdot 1_B|\mathcal{G})$$

2°)

$$E\{1_A \cdot E(||Z||^2 \cdot 1_A|\mathcal{G})\} = E\{E(1_A|\mathcal{G}) \cdot E(||Z||^2 \cdot 1_A|\mathcal{G})\}$$
$$= E\{E(1_B|\mathcal{G}) \cdot E(||Z||^2 \cdot 1_B|\mathcal{G})\} = E\{1_B \cdot E(||Z||^2 \cdot 1_B|\mathcal{G})\}$$

Then, we have also :

$$E\{1_A \cdot E(||Z||^2|\mathcal{G})\}$$
$$= E\{1_A \cdot E(||Z||^2 \cdot 1_A|\mathcal{G})\} + E\{1_A \cdot E(||Z||^2 \cdot 1_B|\mathcal{G})\}$$
$$= E\{1_B \cdot E(||Z||^2 \cdot 1_B|\mathcal{G})\} + E\{1_A \cdot E(||Z||^2 \cdot 1_B|\mathcal{G})\}$$
$$= E(||Z||^2 \cdot 1_B)$$

B-3. Lemma (if only one jump) . Let H be a Banach space. Let u and v be two stopping times, v being predictable. Let S be an H-valued integrable random variable which is $\mathcal{F}_v$ -measurable. We put $C=S-E\{S|\mathcal{F}_{v-}\}$.

Let M be the cadlag martingale defined by $M_t = E(C|\mathcal{F}_t)$. Then M has a jump on the stopping time v and is "fixed" elsewhere. Moreower there exists an H-valued square integrable cadlag martingale W with the following properties:

(i) $W \cdot 1_{[o,u[} = M \cdot 1_{[o,u[}$
(this implies $[W] \cdot 1_{[o,u[} = [M] \cdot 1_{[o,u[}$),

(ii) the random measures $d<W>$ and $d<M>$ are such that $d<W> \leqslant d<M>$

(iii) for each predictable real positive process Y :

$$E\{\int_{]o,u]} Y \cdot d[W]\} = E\{\int_{]o,u]} Y \cdot d<W>\} \leqslant$$

$$E\{\int_{]o,u[} Y \cdot (d[W] + d<W>)\}$$

<u>Proof</u>. 1°) Let $(v(k))_{k>0}$ be a sequence of stopping times "announcing" v (i.e. $v(k) \uparrow v$ and, $\forall k, P([v(k) < v]) = 1$).
For each integer k, $E(C | \mathcal{F}_{v(k)}) = 0$, then

$$M.1_{[0,v(k)]} = 0 ;$$

this implies $M.1_{[0,u[} = 0$. Moreover, C being $\mathcal{F}_v$-measurable, $M_1 = M_v$. This proves that M has a jump on the stopping time v and is "fixed" elsewhere.

2°) For the building of W, we can suppose that M is stopped at u (i.e. $M_1 = M_u$). Then we consider the sets $B = \{\omega : v(\omega) = u(\omega)\}$ and $A = \Omega \setminus B$, the σ-algebra $\mathcal{G} = \mathcal{F}_{v-}$, the σ-algebra $\mathcal{G}^*$ generated by $\mathcal{G}$ and by the set B, the random variable

$$D_1 = C \cdot 1_B - E(C.1_B | \mathcal{G}^*) = 1_B \cdot \{C - E(C | \mathcal{G}^*)\}$$

and the cadlag martingale D defined by $D_t = E(D_1 | \mathcal{F}_t)$. We note that D has a jump on the stopping time v and is "fixed" elsewhere (cf. 1°) above).

Now we put $W = M - D$ and we shall prove the properties (ii) and (iii) (the property (i) is proved above). We note that $W_v = W_1 = 1_A \cdot C + 1_B \cdot E(C|\mathcal{G}^*)$.

3°) The stopping time v being predictable, we have :

$$< W >_v = < W >_v - < W >_{v-}$$

$$= E([W]_v | \mathcal{G})$$

$$= E(\{ 1_A \cdot ||C||^2 + 1_B \cdot ||E(C| \mathcal{G}^*)||^2 \} | \mathcal{G}),$$

$$\leq E(\{ 1_A \cdot ||C||^2 + 1_B \cdot ||C||^2 \} | \mathcal{G}),$$

$$\leq E([M]_v | \mathcal{G}) \leq < M >_v$$

Actually, we have $d < W > \leq d < M >$, i.e. the property (ii).

4°) Let Y be a predictable real positive process. Then, the random variable Y_v is $\mathcal{F}_{v-}$-measurable. We have :

$$E(\int_{]0,u]} Y \cdot d[W])$$

$$= E(Y_v \cdot 1_A \cdot ||C||^2) + E(Y_v \cdot 1_B \cdot ||E(C| \mathcal{G}^*)||^2)$$

The fisrt term is bounded by $E(\int_{]o,u[} Y \cdot d[W])$.

By the lemma B-2, if we put $Z = (Y_v)^{1/2} \cdot E(C \mid \mathcal{G}^*)$, the second term is equal to

$$E(Y_v \cdot 1_A \cdot E\{||E(C|\mathcal{G}^*)||^2 \mid \mathcal{G}\})$$

which is bounded by $E(\int_{]o,u[} Y \cdot d<W>)$ (see above).

This proves the property (iii).

B-4. Proposition. Let H be a Hilbert space. Let M be an H-valued cadlag square integrable martingale. Let u be a stopping time. Then, there exists an H-valued cadlag square integrable martingale W with the following properties :

(i) $W.1_{[o,u[} = M.1_{[o,u[}$ (this implies :

$$[W].1_{[o,u[} = [M].1_{[o,u[}$$

(ii) the "random measures" $d<W>$ and $d<M>$ are such that $d<W> \leqslant d<M>$.

(iii) for each predictable real positive process :

$$E\{\int_{]o,u]} Y.d[W]\} = E\{\int_{]o,u]} Y.d<W>\}$$

$$\leqslant E\{\int_{]o,u[} Y.(d[W] + d<W>\}$$

$$\leqslant E\{\int_{]o,u[} Y \cdot d[M] + d<M>\}$$

Proof. We can assume that $M_1 = M_u$. Let w be a "totally inacessible" stopping time and let $(v(n))_{n>0}$ be a sequence of predictable stopping times such that $[u] \subset [w] \cup \{\bigcup_{n>0} [v(n)]\}$, we can assume that the sets $[v(n)]_{n>0}$ are disjoint

For each integer n we define the random variable $C_n = M_{v(n)} - M_{v(n)-}$ and $M_t^n = E(C_n \mid \mathcal{F}_t)$.

We can define $\overline{M} = M - \sum_{n>0} M^n$ (convergent series in the space of square integrable martingales) and we have $[M] = [\overline{M}] + \sum_{n>0} [M^n]$ (the sets $[v(n)]_{n>0}$ being disjoint). Moreover, $<\overline{M}>$ being a

predictable process, $<\bar{M}>_w = <\bar{M}>_{w-}$; this implies $<\bar{M}>_u = <\bar{M}>_{u-}$
Then, for each predictable real positive process Y, we have :

$$E(\int_{]o,u]} Y \cdot d<\bar{M}> = E(\int_{]o,u]} Y \cdot d<\bar{M}>)$$

$$= E(\int_{]o,u[} Y \cdot d<\bar{M}>)$$

(if we define $\bar{W} = \bar{M}$, the properties (i), (ii) and (iii) are satisfied for the pair $(\bar{W},\bar{M})$).
Then, for each integer n, we build a martingale W^n associated to M^n as in the lemma B-3. By additivity, the proposition is proved if we put

$$W = \bar{W} + \sum_{n>0} W^n = \bar{M} + \sum_{n>0} W^n$$

B-5. <u>Corollary</u>. Let M be a Hilbert space valued square integrable cadlag martingale. Then, for each stopping u and for each real bounded predictable process Y, we have:

$$E(\sup_{o\leq t<u} ||\int_{]o,t]} Y_s \cdot dZ_s||^2)$$

$$\leq 4 \cdot (E\{\int_{]o,u[} Y_s^2 \cdot (d<M>_s + d[M]_s)\})$$

(cf. A-3.e) for the notation above).

<u>Proof.</u> Let W be a martingale associated to M and u as in the proposition B-4 above. The stochastic integral $\int_{]o,u]} Y_s . dW_s$ is well defined (see the property (ii)) and we have :

$$(\int Y \cdot dM.1_{]o,u[} = (\int Y \cdot dW)1_{]o,u[}$$

(this is obwious if Y is an $\mathcal{A}$-simple process, and it is true in the general case by linearity and density).

Then, we have :

$$E\{\sup_{o\leq t<u} ||\int_{]o,t]} Y_s \cdot dM_s||^2)$$

$$= E\{\sup_{o\leq t<u} ||\int_{]o,t]} Y_s \cdot dW_s||^2)$$

$$\leq E\{\sup_{o\leq t<u} ||\int_{]o,t]} Y_s \cdot dW_s ||^2)$$

$$\leq 4 \cdot E\left(\left|\left| \int_{]o,u]} Y_s \cdot dW_s \right|\right|^2 \right) \quad \text{(Doob inequality)}$$

$$\leq 4 \cdot E\left(\int_{]o,u]} Y_s^2 \cdot d[W] \right)$$

$$\leq 4 \cdot E\left(\int_{]o,u[} Y_d^2 \cdot \{d[M] + d <M>\} \right) \text{ (cf. B-4).}$$

B-6. Theorem. We consider a Banach space H, two Hilbert spaces J and K, and a bilinear mapping of $H \times J$ into K which, to (x,y) element of $(H \times J)$, associates $y.x$ element of K. We suppose that we have, for each element $(y.x)$ of $(H \times J)$,

$$||y.x|| \leq ||y||.||x||.$$

Let M be a J-valued cadlag locally square integrable martingale. Let V be a J-valued cadlag adapted process of finite variation (i.e., for each element ω of Ω, the function $s \rightsquigarrow V_s(\omega)$ has a bounded variation). We put $Z = M + V$. Then, there exists an increasing cadlag adapted process A such that,

(i) for each stopping time u and for each H-valued predictable process Y :

$$E\{ \sup_{t<u} \left|\left| \int_{]o,t]} Y \cdot dZ \right|\right|_K^2 \leq E(A_{u-}\{\int_{]o,u[} ||Y_t||^2 . dA_t\})$$

(see A-3-e) for the notation above).

Moreover, if the dimension of J is finite, we have the same property if M is a J-valued cadlag local martingale (i.e. Z is a semi-martingale).

Proof. 1°) The set of processes Z for which there exists a process A with the property (i) is clearly a vector space. Moreover, if Z = V, is a process of finite variation $B_t = \int_{]o,t]} d||V_s||$ and if we put $A_t = B_t$, the condition (i) is satisfied by the Cauchy-Schwartz inequality (applied for each "sample function"). Then, it sufficient to prove the theorem when Z = M is a locally square integrable martingale. In this case, the condition (i) is satisfied if we put $A_t = (<M>_t + [M]_t) + 1$ (cf. the corollary B-5).

2°) When the dimension of J is finite, each local martingale is the sum of a process of finite variation and a locally square integrable martingale (cf. [3]) and we can apply the previous result.

References

[1] C. Doleans-Dade : On the existence and unicity of stochastic integral equations.

[2] M. Metivier et J. Pellaumail : Inégalites pour une martingale et équations différentielles stochastiques. Séminaire de Rennes 1977 (to appear).

[3] P.A. Meyer : Séminaire de probabilités X, Springer-Verlag 1976.

I.N.S.A. - B.P. 14, A,
35031 - Rennes Cedex
F r a n c e

THE ORDER OF APPROXIMATION IN THE RANDOM CENTRAL LIMIT THEOREM

Z. Rychlik

1. Introduction. Let $\{X_n, n \geq 1\}$ be a sequence of independent random variables such that $EX_k = \alpha_k$, $\sigma^2 X_k = \delta_k^2$ and $E|X_k - \alpha_k|^3 = \beta_k^3 < \infty$.

Let us put

$$S_n = \sum_{k=1}^{n} X_k, \quad s_n^2 = \sum_{k=1}^{n} \delta_k^2, \quad B_n^3 = \sum_{k=1}^{n} \beta_k^3 .$$

Now let $\{N_n, n \geq 1\}$ be a sequence of positive integer-valued random variables. Define

$$S_{N_n} = \sum_{k=1}^{N_n} X_k, \quad L_n = \sum_{k=1}^{N_n} \alpha_k, \quad M_n^2 = \sum_{k=1}^{N_n} \delta_k^2$$

Following the classical work of Robbins [15], many authors (see, e.g., [1 - 10], [13 - 17]) have investigated the limit behaviour of the distribution of sums with random indices. Such investigations are not only of theoretical interest but are also important in various applications, e.g. in the theory of Markov chains, in sequential analysis, in random walk problems, in connection with Monte Carlo Methods and in the theory of queues [8].

The case when N_n is independent of the X_k-s has been treated by Robbins [15], Gnedenko, Fahim [7], Szász, Freyer [17], and the author of the present note [16]. The first essential result in the case when N_n is not assumed to be independent of the X_k-s belongs to Anscombe [1]. The generalizations of Anscombe's result have been given by Rényi [14] Mogyoródi [13], Blum, Hanson, Rosenblatt [4], Guiasu [9], and Csörgő, Fischler [6].

On the other hand, those results have established the background for studying the problem of weak convergence of randomly selected partial sum process in function spaces, which give much more information. These problems began with Billingsley [3], and were extended by M. Csörgő, S. Csörgő [5], Babu and Gosh [2].

From the main result of [2] we can derive the following generalization of the result given in [4].

Theorem 1. Let $\{X_n, n \geq 1\}$ be a sequence of independent random variables such that $EX_k = 0$, $EX_k^2 = \delta_k^2 < \infty$. Assume that the Lindeberg condition is satisfied

$$s_n^{-2} \sum_{k=1}^{n} EX_k^2 I(|X_k| \geq \varepsilon s_n) \longrightarrow 0, \text{ as } n \longrightarrow \infty, \text{ for all}$$

$\varepsilon > 0$.

If there exists a sequence $\{k_n, n \geq 1\}$ of positive integers such that $k_n \longrightarrow \infty$ as $n \longrightarrow \infty$, and $M_n^2/s_{k_n}^2 \xrightarrow{P} \lambda$, for some positive random variable λ, then

(1) $$\lim_{n\to\infty} P[S_{N_n} - L_n < xM_n] = \Phi(x)$$

or equivalently

(2) $$\lim_{n\to\infty} P[S_{N_n} - L_n < x\, s_{k_n} \lambda^{1/2}] = \Phi(x).$$

The aim of this paper is to give, under additional assumptions, the rate of convergence in (1) and (2). In the case when $\{X_n, n\geq 1\}$ is a sequence of independent and identically distributed random variables our results reduce to the results given in [2].

2. The results. We write $d_n = O(b_n)$ if and only if $|d_n| \leq C|b_n|$ for all $n \geq 1$ with some appropriate constant $C > 0$. Hereafter C will denote absolute constants, and the same symbol may be used for different constants.

Theorem 2. Let $\{X_n, n \geq 1\}$ be a sequence of independent random variables such that $EX_k = \alpha_k$, $\sigma^2 X_k = \delta_k^2$, $E|X_k - \alpha_k|^3 = \beta_k^3 < \infty$, and let $\{N_n, n \geq 1\}$ be a sequence of positive integer-valued random variables. If $B_n^3/s_n^3 \leq \sqrt{\varepsilon_n}$ and

(3) $$P[|M_n^2/s_n^2 \lambda - 1| \geq \varepsilon_n] = O(\sqrt{\varepsilon_n})$$

for some constant $\lambda > 0$ and a sequence $\{\varepsilon_n, n \geq 1\}$, $\varepsilon_n \longrightarrow 0$ as $n \longrightarrow \infty$, then

(4) $$\sup_x |P[S_{N_n} - L_n < x\, s_n \lambda^{1/2}] - \Phi(x)| = O(\sqrt{\varepsilon_n})$$

and

(5) $$\sup_x |P[S_{N_n} - L_n < xM_n] - \Phi(x)| = O(\sqrt{\varepsilon_n}).$$

From Theorem 2 we get immediately the main result of [10].

Theorem 3. Let $\{X_n, n \geq 1\}$ be a sequence of independent and identically distributed random variables with $EX_k = \alpha$, $\sigma^2 X_k = \delta^2$ and $\beta_1^3 = E|X_1 - \alpha|^3 < \infty$. Let $\{N_n, n \geq 1\}$ be a sequence of positive integer-valued random variables. If

$$P[|N_n/n\lambda - 1| \geq \varepsilon_n] = O(\sqrt{\varepsilon_n})$$

for some constant $\lambda > 0$ and a sequence ε_n with $1/n \leq \varepsilon_n \longrightarrow 0$ as $n \longrightarrow \infty$, then

$$\sup_x |P[S_{N_n} - \alpha N_n < x \delta\sqrt{\lambda n}] - \Phi(x)| = O(\sqrt{\varepsilon_n})$$

and

$$\sup_x |P[S_{N_n} - \alpha N_n < x \delta\sqrt{N_n}] - \Phi(x)| = O(\sqrt{\varepsilon_n}).$$

Proof of Theorem 2. Without los of generality we may assume that $EX_k = 0$, $k = 1, 2, \ldots$. Let $b_n(x) = x s_n\sqrt{\lambda}$ and let $I_n = \{k \geq 1 : s_n^2 \lambda(1-\varepsilon_n) \leq s_k^2 \leq s_n^2\lambda(1+\varepsilon_n)\}$. Then, according to (3),

(6) $$P[N_n \notin I_n] = O(\sqrt{\varepsilon_n})$$

Thus, to prove (4), it suffices to show

$$\sup_x |P[S_{N_n} < b_n(x) \; ; \; N_n \in I_n] - \Phi(x)| = O(\sqrt{\varepsilon_n}).$$

Define

$$A_n(x) = [\max_{k\in I_n} S_k < b_n(x)], \quad B_n(x) = [\min_{k\in I_n} S_k < b_n(x)].$$

Then, for every real number x, we get

(7) $$P[A_n(x) \cap (N_n \in I_n)] \leq P[S_{N_n} < b_n(x); \; N_n \in I_n] \leq P[B_n(x)].$$

On the other hand, we have

(8) $$P[A_n(x)] \leq P[\sum_{k\in D_n} X_k < b_n(x)] \leq P[B_n(x)],$$

where $D_n = \{k \geq 1 : s_k^2 \leq \lambda s_n^2\}$. Moreover, the Theorem of Berry-Esseen [11, p. 288] shows that

(9) $$\sup_x \left| P\left[\sum_{k \in D_n} X_k < b_n(x)\right] - \Phi(x)\right| = O(B_n^3/s_n^3)$$

and in view of our assumption, $B_n^3/s_n^3 = O(\sqrt{\varepsilon_n})$. Hence, according to (6), (7), (8) and (9).

$$\sup_x | P[S_{N_n} < b_n(x) ; N_n \in I_n] - \Phi(x)| \leq O(\sqrt{\varepsilon_n}) +$$

$$+ \sup_x (P[B_n(x)] - P[A_n(x)]) .$$

Thus, in order to prove (4), it suffices to show that

(10) $$\sup_x (P[B_n(x)] - P[A_n(x)]) = O(\sqrt{\varepsilon_n}).$$

Now let us put

$$E_n = \{k \geq 1: s_k^2 \leq \lambda s_n^2(1 - \varepsilon_n)\} ,$$

$$F_n = \{k \geq 1: s_k^2 \leq s_n^2 \lambda(1 + \varepsilon_n)\} .$$

Then, according to Lemma 1, there exists a constant C which is independent of x and n, such that

(11) $$P[B_n(x) - P[A_n(x)] \leq C \{ P[\sum_{k \in E_n} X_k \leq b_n(x) ;$$
$$\sum_{k \in F_n} X_k \geq b_n(x)] + P[\sum_{k \in E_n} X_k \geq b_n(x) ; \sum_{k \in F_n} X_k \leq b_n(x)]\}$$

On the other hand, according to Lemma 2, there exists a constant C such that

(12) $$\sup_x P[\sum_{k \in E_n} X_k \leq b_n(x) ; \sum_{k \in F_n} X_k \geq b_n(x)] \leq O(\sqrt{\varepsilon_n}) +$$

$$+ C[s_n^2 \lambda(1+\varepsilon_n) - s_n^2 \lambda(1-\varepsilon_n)]^{1/2}/[s_n^2\lambda(1-\varepsilon_n)]^{1/2} = O(\sqrt{\varepsilon_n}),$$

and

$$(13)\qquad \sup_x P[\sum_{k\in E_n} X_k \geq b_n(x)\ ;\ \sum_{k\in F_n} X_k \leq b_n(x)] \leq O(\sqrt{\varepsilon_n}) +$$

$$+ C[s_n^2\lambda\,(1+\varepsilon_n) - s_n^2\lambda(1-\varepsilon_n)]^{1/2}/[s_n^2\,\lambda(1-\varepsilon_n)]^{1/2} = O(\sqrt{\varepsilon_n})$$

Therefore (10) holds and the proof of (4) is complete. To prove (5) suffices to observe that

$$[\ |M_n/s_n\,\lambda^{1/2} - 1\ | \geq \sqrt{\varepsilon_n}] \subset [|\ M_n^2/s_n^2\lambda - 1| \geq \varepsilon_n].$$

Hence, by (3) ,

$$P\,[|\ M_n/s_n\lambda^{1/2} - 1| \geq \sqrt{\varepsilon_n}] = O(\sqrt{\varepsilon_n}),$$

and therefore (5) follows from (4) and Lemma 3.

Remark. Taking into account the Remarks and the Examples given in the paper (10), one can observe that in general the rate of convergence in (4) or (5) cannot be sharpened, even if the assumption (3) is maximally sharpened to

$$P[|\ M_n^2/s_n^2\,\lambda - 1\ | \geq \varepsilon_n] = 0.$$

Furthermore, in general the order of approximation cannet be improved by additional assumptions on the dependence of the random variables $\{N_n,\ n \geq 1\}$ and the sequence $\{X_n,\ n \geq 1\}$. Such assumptions, which are very important in applications, are for example, the event $[N_n < k]$ is independent of $X_j, j \geq k$, or $N_n, n \geq 1$, are stopping times with respect to $\{X_k,\ k \geq 1\}$. Thus in the dependent case stronger assumption on $\{M_n^2, n \geq 1\}$ does not lead beyond the approximation order $O(\sqrt{\varepsilon_n})$. But in the case when $N_n, n \geq 1$, are independent of the sequence $\{X_n, n \geq 1\}$ the sharpening of the assumption (3) can lead to sharpening of the approximation order in (4) or (5) what follows from the following Theorem.

<u>Theorem</u> 4. Let $\{X_n, n \geq 1\}$ be a sequence of independent random variables with $EX_k = \alpha_k$, $\sigma^2 X_k = \delta_k^2$, $E|X_k - \alpha_k|^3 = \beta_k^3 < \infty$, and let, for each $n \geq 1$, the random variable N_n be independent of the sequence $\{X_k, k \geq 1$. If $B_n^3/s_n^3 \leq \varepsilon_n$ and

(14) $P[| M_n^2/s_n^2 \lambda - 1| \geq \varepsilon_n] = O(\varepsilon_n)$

for some constant $\lambda > 0$ and a sequence $\{\varepsilon_n, n \geq 1\}$, $\varepsilon_n \longrightarrow 0$ as $n \longrightarrow \infty$, then

(15) $\sup_x |P[S_{N_n} - L_n < x\, s_n \lambda^{1/2}] - \Phi(x) = O(\varepsilon_n)$.

If

(16) $P[1- M_n^2/s_n^2\lambda \geq \alpha] = O(\varepsilon_n)$

for some $\lambda > 0$ and $\alpha < 1$, then

(17) $\sup_x |P[S_{N_n} - L_n < xM_n] - \Phi(x)| = O(\varepsilon_n)$.

Proof. Let, as above, $I_n = \{k \geq 1 : s_n^2\lambda(1-\varepsilon_n) \leq s_k^2 \leq s_n^2\lambda(1-\varepsilon_n)\}$. We assume that $EX_k = 0$, $k \geq 1$. Then, by (14) and by the independence N_n of $\{X_k, k \geq 1\}$, we get

(18) $\sup | P[S_{N_n} - L_n < x\, s_n\lambda^{1/2}] - \Phi(x)| \leq$

$$\sum_{k\in I_n} P[N_n = k] \sup_x |P[S_k/s_k < x\, s_n\lambda^{1/2}/s_k] - \Phi(x)|$$

$$+ \sum_{k\in I_n} P[N_n = k] \sup_x |\Phi(xs_n\lambda^{1/2}/s_k) - \Phi(x)| + O(\varepsilon_n).$$

On the other hand, by the Berry-Esseen inequality, we have

(19) $\sup_{k\in I_n} \sup_x |P[S_k/s_k < xs_n \lambda^{1/2}/s_k] - \Phi(x\, s_n \lambda^{1/2}/s_k)|$

$$\leq C \sup_{k\in I_n} (B_k^3/s_k^3) = O(\varepsilon_n).$$

Furthermore, for all x and for every $k \in I_n$, we hawe

(20) $| \Phi(x\, s_n \lambda^{1/2}/s_k) - \Phi(x)| \leq C\, \varepsilon_n$.

Thus, by (18),(19) and (20), we obtain the assertion (15).

The second part of this Theorem runs similarly. Namely, putting $I_n' = \{ k \geq 1:\ s_k^2 \geq s_n^2 \lambda(1 -\alpha)\}$, we get

$$\sup_x |P[S_{N_n} < xM_n] - \Phi(x)| \leq O(\varepsilon_n) +$$

$$\sum_{k\in I'_n} P[N_n = k] \ \sup_x | P[S_k < xs_k] - \Phi(x)| .$$

Since by the Barry-Esseen inequality

$$\sup_{k\in I'_n} \sup_x |P[S_k < x s_k] - \Phi(x)| \leq C \sup_{k\in I'_n} (B_k^3/s_k^3) = O(\varepsilon_n),$$

the assertion (17) follows, and hence the proof is finished.

3. Auxiliary Lemmas. In this Section we state and prove some Lemmas which were needed for the proof of our results. Lemmas 1 and 2 are extensions, to non-identically distributed random variables, of the Lemmas 7 and 8 given in (10), while Lemma 3 is only cited from (12) for the sake of completeness.

Lemma 1. Let $\{X_n, n \geq 1\}$ be a sequence of independent random variables such that $EX_k = 0$, $EX_k^2 = \sigma_k^2 < \infty$. If

$$\lim_{n\to\infty} s_n^{-2} \sum_{k=1}^{n} EX_k^2 I(|X_k| \geq \varepsilon s_n) = 0$$

for all $\varepsilon > 0$, then there exists a constant C such that for any positive numbers p and q, $p < q$, and for every x

$$P[\min_{k\in H(p,q)} S_k < x] - P[\max_{k\in H(p,q)} S_k < x) \leq$$

$$C \{P [\sum_{k\in H(0,p)} X_k \leq x; \sum_{k\in H(0,q)} X_k \geq x] +$$

$$+ P [\sum_{k\in H(0,p)} X_k \geq x ; \sum_{k\in H(0,q)} X_k \leq x]\},$$

where $H(r,s)$ denotes the set $\{ k \geq 1 : r \leq s_k^2 \leq s \}$.

Proof. Let p,q and x be given. We must show that for an appropriate constant C, not depending on p,q and x,

$$P(A) \leq C(P_1 + P_2) ,$$

where

$$A = [S_j \leq x;\ \ S_k > x \ \text{ for some } j, k \text{ with } \ p \leq s_j^2,\ s_k^2 \leq q\ ],$$

$$P_1 = P\,[\sum_{k\in H(0,p)} X_k \leq x,\ \ \sum_{k\in H(0,q)} X_k \geq x\ ],$$

$$P_2 = P\,[\sum_{k\in H(0,p)} X_k \geq x,\ \ \sum_{k\in H(0,q)} X_k \leq x\ ].$$

But

$$(21)\qquad P(A) = P(A \cap [\sum_{k\in H(0,p)} X_k \leq x]) + P(A \cap [\sum_{k\in H(0,p)} X_k > x])$$

$$= I_1 + I_2.$$

Thus it suffices to prove $I_1 \leq CP_1$, $I_2 \leq CP_2$. We shall prove only the first inequality, as the proof of the second one has exactly the same pattern.

We have

$$(22)\qquad I_1 = P(A \cap [\sum_{k\in H(0,p)} X_k \leq x] \cap [\sum_{k\in H(0,q)} X_k \leq x])$$

$$+ P(A \cap [\sum_{k\in H(0,p)} X_k \leq x] \cap [\sum_{k\in H(0,q)} X_k > x]).$$

Therefore, taking into account (22) and the definition of P_1 it suffices to prove

$$(23)\qquad P(A \cap [\sum_{k\in H(0,p)} X_k \leq x] \cap [\sum_{k\in H(0,q)} X_k \leq x]) \leq CP_1.$$

On the other hand, it follows from the Lindenberg-Feller central limit theorwm [11, p. 280]

$$\lim_{n\to\infty} P[S_n \leq 0] = \lim_{n\to\infty} P[S_n \geq 0] = 1/2.$$

Thus for all $n \geq 1$

$$(24)\qquad P[S_n \leq 0] \leq CP[S_n \geq 0],$$

where C is some appropriate constant.

Now let us put $G(p,k) = \{j \geq 1:\ p \leq s_j^2 \leq s_{k-1}^2\}$, $A_k = [S_j \leq x$ for $j \in G(p,k)$ and $S_k > x]$, $h(p,q) = \{k \geq 1 :$

$p < s_k^2 < q\}$. Then, by the definition of the sets A_k, and by (24), we get

$$P(A \cap [\sum_{k\in H(0,p)} X_k \leq x] \cap [\sum_{i\in H(0,q)} X_i \leq x]) =$$

$$= \sum_{k\in h(p,q)} P(A_k \cap [\sum_{k<i\in H(0,q)} X_i \leq x]) \leq$$

$$\leq \sum_{k\in h(p,q)} P(A_k \cap [\sum_{k<i\in H(0,q)} X_i \leq 0]) =$$

$$= \sum_{k\in h(p,q)} P[A_k]\, P[\sum_{k<i\in H(0,q)} X_i \leq 0] \leq$$

$$\leq C \sum_{k\in h(p,q)} P[A_k]\, P[\sum_{k<i\in H(0,q)} X_i \geq 0] =$$

$$= C \sum_{k\in h(p,q)} P(A_k \cap [\sum_{i\in H(0,q)} X_i \geq 0]) = CP_1.$$

Therefore (23) holds and the proof of the Lemma is complete.

Lemma 2. Let $\{X_n,\ n \geq 1\}$ be a sequence of independent random variables with $EX_k = 0$, $EX_k^2 = \delta_k^2$, $\beta_k^3 = E|X_k|^3 < \infty$, $k \geq 1$. Then there exists a constant C such that for all $k,\ n \geq 1$, and every x,

$$P[S_n \leq x,\ S_{n+k} \geq x] \leq C\{B_n^3/s_n^3 + (s_{n+k}^2 - s_n^2)^{1/2}/s_n\}, \tag{25}$$

and

$$P[S_n \geq x\ ,\ S_{n+k} \leq x] \leq C\{B_n^3/s_n^3 + (s_{n+k}^2 - s_n^2)^{1/2}/s_n\} \tag{26}$$

Proof. It is enough to prove (25) since (26) follows from (25) by replacing X_k by $-X_k$.

Let P_{X_n} be the distribution of X_n, and let $*$ denote the product of measures. Then using Fubini's theorem, we get

$$P[S_n \leq x,\ S_{n+k} \geq x] = P_{X_1} * P_{X_2} * \ldots * P_{X_{n+k}} \{(z_1, z_2, \ldots, z_{n+k}) :$$

$$\sum_{i=1}^{n} z_i \leq x,\ \sum_{i=1}^{n+k} z_i \geq x\} = P_{X_1} * P_{X_2} * \ldots * P_{X_{n+k}} \{(z_1, \ldots, z_{n+k}) :$$

$$(\sum_{i=1}^{n} z_i)/s_n \leq x/s_n,\ (\sum_{i=1}^{n} z_i)/s_n + (\sum_{i=n+1}^{n+k} z_i)/s_n \geq x/s_n\} =$$

$$= \int P_{X_1} * P_{X_2} * \ldots * P_{X_n} \{(z_1, z_2, \ldots, z_n) : x/s_n - (\sum_{i=n+1}^{n+k} z_i)/s_n$$

$$\leq (\sum_{i=1}^{n} z_i)/s_n \leq x/s_n\} \quad P_{X_{n+1}} * P_{X_{n+2}} * \ldots * P_{X_{n+k}} (dz_{n+1}, \ldots, dz_{n+k})$$

and hence, according to the Berry-Esseen's theorem

$$\leq C(B_n^3/s_n^3) + \int |\Phi(x/s_n) - \Phi(x/s_n - \sum_{i=n+1}^{n+k} z_i/s_n) P_{X_{n+1}} *$$

$$* P_{X_{n+2}} * \ldots * P_{X_{n+k}} (dz_{n+1}, dz_{n+2}, \ldots, dz_{n+k}) \leq \{C\ B_n^3/s_n^3$$

$$+ (s_{n+k}^2 - s_n^2)^{1/2})/s_n \int |\sum_{i=n+1}^{n+k} z_i|/(s_{n+k}^2 - s_n^2)^{1/2}\ P_{X_{n+1}} * P_{X_{n+2}} *$$

$$\ldots * P_{X_{n+k}} (dz_{n+1}, \ldots, dz_{n+k}) \leq C\{\ B_n^3/s_n^3 +$$

$$+ (s_{n+k}^2 - s_n^2)^{1/2}/s_n\ \}$$

since

$$E|\sum_{i=n+1}^{n+k} X_i|/(s_{n+k}^2 - s_n^2)^{1/2} \leq 1.$$

Thus establishes Lemma 2.

<u>Lemma</u> 3. [12] Let $\{X_n, n \geq 1\}$ and $\{Y_n, n \geq 1\}$ be two sequences of random variables. Assume that

$$\sup_x |\ P[X_n < x] - \Phi(x)| = O(\varepsilon_n)$$

and

$$P[|Y_n - 1| \geq \varepsilon_n] = O(\varepsilon_n), \text{ then}$$

$$\sup_x |\ P[\ X_n < xY_n] - \Phi(x)| = O(\varepsilon_n).$$

References

[1] F.J. Anscombe, Large-sample theory of sequential estimation, Proc.Cambridge Phil.Soc., 48 (1952),pp. 600-607.

[2] G.J. Babu and M. Ghosh, A random functional central limit theorems for martingales, Acta Math.Acad. Sci.Hung., 27(1976), pp. 301-306.

[3] P.P. Billingsley, Limit theorems for randomly selected partial sums, Ann.Math. Statist., 33(1962), pp. 85-92.

[4] J.R. Blum, D.I. Hanson and J.I. Rosenblatt, On the central limit theorem for the sum of a random number of independent random variables, Z. Wahrscheinlichkeitstheorie verw.Gebiete, 1(1963), pp. 389-393.

[5] M. Csörgö and S. Csörgö, On weak convergence of randomly selected partial sums, Acta Scient. Math., 34(1973),pp.53-60.

[6] M. Csörgo and R. Fischler, Some examples and results in the theory of mixing and random-sum central limit theorems, Periodica Mathematica Hungarica, 3(1973), pp. 41-57.

[7] B.V. Gnedenko and H. Fahim, On a transfer theorem, Dokl. Akad. Nauk SSSR, 187(1969), pp. 15-17.

[8] B.V. Gnedenko, On the relationship of the theory of summation of independent random variables to problems in queueing theory and reliability theory, Revue Roumaine Math. Pures Appl, 12(1967), 1243-1253.

[9] S. Guiasu, On the asymptotic distribution of the sequence of random variables with random indices, Ann. Math.Statist., 42(1971), pp. 2018-2028.

[10] D. Landers and L. Rogge, The exact approximation order in the central-limit-theorem for random summation, Z. Wahrscheinlichkeitstheorie verw.,Gebiete, 36(1976), pp. 269-283.

[11] M. Loeve, Probability Theory, Princeton, N.J., Van Nostrand 1963.

[12] R. Michel and J. Pfanzagl, The accuracy of the normal approximation for minimum contrast estimates, Z. Wahrscheinlichkeitstheorie verw. Gebiete, 18(1971), pp. 73-84.

[13] J. Mogyoródi, A central limit theorem for the sum of a random number of random variables, Publ. Math. Inst.Hungar. Acad. Sci. Ser. A7., (1962), pp. 409-424.

[14] A. Rényi, On the central limit theorem for the sum of a random number of independent random variables, Acta Math. Acad. Sci. Hungar., 11(1960), pp. 92-102.

[15] H. Robbins, The asymptotic distribution of the sums of a random number of random variables, Bull. Amer. Math. Soc., 54(1948), pp. 1151-1161.

[16] Z. Rychlik, A central limit theorem for sums of a random number of independent random variables, Colloquium Mathematicum, 35(1976), pp. 147-158.

[17] D. Szász and B. Freyer, A problem of summation with random indices, Litovskii Mat. Sbornik, 11(1971), pp.181-187.

Institute of Mathematics
Maria Curie-Skłodowska University
20-031 Lublin, Nowotki 10
P o l a n d

BANACH-SPACE VALUED STATIONARY PROCESSES WITH ABSOLUTELY CONTINUOUS SPECTRAL FUNCTION

F. Schmidt

Summary: A generalization of a result of Miamee and Salehi on the factorization of positive operator valued functions on a Banach space is given. Also, two decomposition theorems for Banach-space valued stationary processes are proved, and the connections between the existence of moving averages representations and the regularity of a process on the one side and the absolutely continuity of his spectral function and the factorability of the corresponding density on the other side are investigated. Besides, some regularity conditions are discussed.

Let $\underline{H} = \underline{H}(\Omega) = L^2(\Omega, \mathfrak{A}, \underline{P})$ be the Hilbert space of all (equivalence classes of) complex valued random variables on the probability space $(\Omega, \mathfrak{A}, \underline{P})$ having a finite absolute moment of second order, and let $\underline{B}$ a (complex) Banach space (not necessarily separable). A Banach-space valued stationary process (BSVSP) is a sequence $X = (X_n)_{n \in Z}$ (Z denotes the group of integers) of (bounded linear) operators $X_n \in [\underline{B}, \underline{H}]$ with the property $X_m^* X_n = X_0^* X_{n-m}$ $(m,n \in Z)$. For a BSVSP X let $\underline{H}_X = \bigvee_{n \in Z} (X_n \underline{B})$, $H_X(0) = \bigvee_{n \leq 0} (X_n \underline{B})$. Let U_X be the shift operator on $\underline{H}_X$ induced by X, let $E_X(\cdot)$ be the spectral resolution of U_X, and let $F_X(\cdot)$ be the non-random spectral function of X.

Then we have

$$X_m^* X_n = \int e^{i\mu(m-n)}\, dF_X(\mu) \qquad (m,n \in Z).$$

(Here and in sequel $\int$ stands for $\int_{-\pi}^{+\pi}$)

<u>Theorem 1</u>. Each BSVSP X admits a unique decomposition in the form

$$X_n = X_n^{(1)} + X_n^{(2)} \qquad (n \in Z)$$

where $X^{(i)} = (X_n^{(i)})_{n \in Z}$ $(i = 1,2)$ are BSVSP's with the following properties :

(i) $\underline{H}_X = \underline{H}_{X^{(1)}} + \underline{H}_{X^{(2)}}$

(ii) $\langle f, F_{X^{(1)}}(\cdot)\, g \rangle$ and $\langle f, F_{X^{(2)}}(\cdot)\, g \rangle$ are the absolutely continuous and the singular part of $\langle f, F_X(\cdot)\, g \rangle$, respectively ($f,g \in \underline{B}$).

In the proof we make use of a theorem ([4], §66) which says that

$$\underline{H}_X^{(1)} := \{ h \in \underline{H}_X \mid (E_X(\cdot)\; h, h) \text{ is absolutely continuous} \}$$

and

$$\underline{H}_X^{(2)} := \{ h \in \underline{H}_X \mid (E_X(\cdot)\; h, h) \text{ is singular} \}$$

are closed linear subspace of $\underline{H}_X$ with $\underline{X}_X^{(1)} \oplus \underline{H}_X^{(2)}$ which are invariant with respect to U_X. The BSVSP's $X^{(i)}$ defined by $X_n^{(i)} := P_X^{(i)} X_n$ $(n \in Z)$ where $P_X^{(i)}$ is the orthogonal projection operator from $\underline{H}_X$ to $\underline{H}_X^{(i)}$ $(i = 1,2)$ have the properties described in the theorem.

Let $\underline{\underline{P}}(\underline{B})$ be the set of all mappings

$$w: \underline{B} \times \underline{B} \ni (f,g) \longrightarrow w(\cdot,f,g) \in L^1(-\pi,+\pi)$$

with the properties

(i) $w(\cdot,f,f) \geq 0 \quad (f \in \underline{B})$

(ii) $w(\cdot,f,g)$ is linear in f and is antilinear in g

(iii) $\int w(\mu,f,f)d\mu \leq C\, \|f\|^2 \quad (f \in \underline{B})$ for a finite constant C.

If W is a weakly summable function on $[-\pi, +\pi]$ with values in $[\underline{B},\underline{B}^*]$ with $\langle f, W(\mu)\, f \rangle \geq 0$ $(\mu \in [-\pi,+\pi],\ f \in \underline{B})$ then

(1) $\quad w(\cdot,f,g) := \langle f, W(\cdot)\, g \rangle \quad (f,g \in \underline{B})$

defines a $w \in \underline{\underline{P}}(\underline{B})$. However, this is not the most general form of $w \in \underline{\underline{P}}(\underline{B})$! If X is a BSVSP for which $F_X(\cdot)$ is weakly absolutely continuous (i.e., $\langle f, F_X(\cdot)\, g \rangle$ $(f,g \in \underline{B})$ is absolutely continuous) then w_X belongs to $\underline{\underline{P}}(\underline{B})$ where $w_X(\cdot,f,g)$ is the derivative of $\langle f, F_X(\cdot)\, g \rangle$ $(f,g \in \underline{B})$. Conversely, for every $w \in \underline{\underline{P}}(\underline{B})$ there exists a BSVSP X for which $F_X(\cdot)$ is weakly absolutely continuous such that $w_X = w$, i.e.

$$\underline{E}\, (X_m f)\overline{(X_n g)} = \int e^{i\mu(n-m)}\, w(\mu,f,g)d\mu \quad (m,n \in Z,\ f,g \in \underline{B}).$$

Let $\underline{K}$ be a Hilbert space and let $L^2(\underline{K})$ be the Hilbert space of all strongly measurable functions $\underline{v} : [-\pi,+\pi] \longrightarrow \underline{K}$ for which $\|\underline{v}(\cdot)\|^2$ is summable (w.r.t. Lebesgue measure) and let $H^2(\underline{K})$

be the corresponding Hardy-class consisting of all functions in $L^2(\underline{K})$ with an analytic Fourier expansion. It is easy to see that for every (bounded linear) operator $A^{\sim} \in [\underline{B}, L^2(\underline{K})]$ the mapping w defined by

(2) $w(\cdot, f, g) := (1/2\pi) \; ((A^{\sim}f)(\cdot), (A^{\sim}g)(\cdot)) \quad (f, g \in \underline{B})$

belongs to $\underline{\underline{P}}(B)$. Conversely, we can prove the following

<u>Theorem 2.</u> Let $w \in \underline{\underline{P}}(\underline{B})$. Then there exist a Hilbert space $\underline{K}$ and an operator $A^{\sim} \in [\underline{B}, L^2(\underline{K})]$ such that (2) holds,

<u>Proof.</u> Let X be a BSVSP with $w_X = w$. Then $(E_X(\cdot)h, h)$ is absolutely continuous for all $h \in \underline{H}_X$. By [5] (21,14) there exist $h_\gamma \in \underline{H}_X$ $(\gamma \in \Gamma)$ such that $\underline{H}_X = \bigoplus_{\gamma \in \Gamma} \bigvee_{n \in Z} \{U_X^n h_\gamma\}$. Let $x_n(\gamma) := U_X^n h_\gamma$ $(n \in Z, \gamma \in \Gamma)$.
Then the stationary process $x(\gamma) = (x_n(\gamma))_{n \in Z}$ $(\gamma \in \Gamma)$ has the absolutely continuous spectral function $(E_X(\cdot)h_\gamma, h_\gamma)$ $(\gamma \in \Gamma)$. By a result of Kolmogorov [6] each $x(\gamma)$ $(\gamma \in \Gamma)$ admits a representation in moving averages. From these representations we can construct a moving average representation for X. The conclusion of the theorem follows by [10], Satz 2.1,2.
If we use the generalization of Kolmogorov's result for stationary processes on arbitrary LCA groups proved by Blum/Eisenberg [1] we can show that Theorem 2 holds also in this case.
By application of Theorem 2 for $w(\cdot, f, g) := \langle f, W(\cdot) g \rangle$ $(f, g \in \underline{B})$ we obtain the following result of Miamee and Salehi [9]:

<u>Corollary.</u> Let $\underline{B}$ be a separable Banach space and W be a weakly summable function on $[-\pi, +\pi]$ with values in $[\underline{B}, \underline{B}^*]$ with $\langle f, W(\mu) f \rangle \geq 0$ $(\mu \in [-\pi, +\pi], f \in \underline{B})$. Then there exist a Hilbert space $\underline{K}$ and a strongly measurable function Q on $[-\pi, +\pi]$ with values in $[\underline{B}, \underline{K}]$ such that

(3) $W(\cdot) = Q(e^{i\cdot})^* \, Q(e^{i\cdot})$

holds.
Let $Y = (Y_n)_{n \in Z}$ be a "fundamental" BSVSP with values in $\underline{K}$ (i.e., $Y_n \in [K, H]$ $(n \in Z)$ and $Y_m^* Y_n = I_{\underline{K}}$ for m = n and = 0 for $m \neq n$). Let $(A_k)_{k \in Z}$ be a sequence of operators $A_k \in [\underline{B}, \underline{K}]$ with $\sum_{k=-\infty}^{+\infty} ||A_k f||^2 < \infty$ $(f \in \underline{B})$. Then the "moving averages"

$$X_n := \sum_{k=-\infty}^{+\infty} Y_k A_{n-k} \qquad (n \in Z)$$

define a BSVSP X with values in $\underline{B}$. If in addition $A_{-1} = A_{-2} = \dots = 0$, then we have a "one-sided moving average".

Combining [10], Satz 2.1.2, with the proof of Theorem 2 we can show the

<u>Theorem 3.</u> The BSVSP X has a representation in form of moving averages if and only if $F_X(\cdot)$ is weakly absolutely continuous.

The mapping w $\underline{\underline{P}}(\underline{B})$ is called "factorable" if the Hilbert space $\underline{K}$ and the operator $A^{\sim}$ in Theorem 2 can be chosen such that $A^{\sim}\underline{B} \subset H^2(\underline{K})$. If $\underline{B}$ is separable and w has the form (1), then w is factorable if and only if there exist a Hilbert space $\underline{K}$ and a strongly measurable function Q on $[-\pi,+\pi]$ with values in $[\underline{B},\underline{K}]$ for which $Q(e^{i\cdot})f \in H^2(\underline{K})$ $(f \in \underline{B})$ such that (3) holds.

<u>Theorem 4.</u> Let X be a BSVSP. Then the following conditions are equivalent :

(i) X is regular.

(ii) X has a representation in form of one-sided moving averages.

(iii) $F_X(\cdot)$ is weakly absolutely continuous and w_X is factorable.

<u>Proof.</u> (i) ==> (ii) and (ii) ==> (iii) are well known (see, f.i. [10]). (iii) ==> (i) : Since $\underline{E}\,(X_m f)\overline{(X_n g)} = \int e^{i\mu(n-m)} w_X(\mu,f,g)\, d\mu = (1/2\pi) \int (e^{-i\mu m}(A^{\sim} f)(\mu),\ e^{-i\mu n}(A^{\sim} g)(\mu))\, d\mu$ $(m,n \in Z,\ f,g \in \underline{B})$ there exists an isometric operator V from $\underline{H}_X$ to $L^2(\underline{K})$ such that $VX_m f = (1/2\pi)^{1/2}\ e^{-i.m} A^{\sim} f$ $(m \in Z,\ f \in \underline{B})$ holds. Obviously, $V\,\underline{H}_X(0) \subseteq H^2(\underline{K})$ and $VU_X^n h = e^{-i.n}\, V h$ $(h \in \underline{H}_X)$. For $h \in \underline{H}_X^-(:= \bigcap_{n \in Z} U_X^n\, \underline{H}_X(0))$ we have $V h \in \bigcap_{n \in Z} e^{-i.n} H^2(\underline{K}) = \{0\}$, i.e. $\underline{E}|h|^2 = ||Vh||^2 = 0$, i.e., X is regular.

<u>Theorem 5.</u> Each BSVSP X admits a unique decomposition in the form

$$X_n = X'_n + X''_n + X'''_n \qquad (n \in Z)$$

where $X' = (X'_n)_{n \in Z}$, $X'' = (X''_n)_{n \in Z}$ and $X''' = (X'''_n)_{n \in Z}$ are BSVSP's with the following properties :

(i) $\underline{H}_X = \underline{H}_{X'} + \underline{H}_{X''} + \underline{H}_{X'''}$.

(ii) X' is regular, X'' and X''' are singular.

(iii) $< f, F_{X'}(\cdot)\, g >$ and $< f, F_{X''}(\cdot)\, g >$ are absolutely continuous, $< f, F_{X'''}(\cdot)\, g >$ is singular ($f, g \in \underline{B}$).

<u>Proof</u>. Let $X_n^{(i)} = X_{n,+}^{(i)} + X_{n,-}^{(i)}$ ($n \in Z$, $i = 1,2$) be the Wold decomposition for the process $X^{(i)}$ of Theorem 1 ($X_+^{(i)} = (X_{n,+}^{(i)})_{n \in Z}$ is a regular BSVSP, $X_-^{(i)} = (X_{n,-}^{(i)})_{n \in Z}$ is a singular BSVSP). By Theorem 4, the functions $F_{X_+^{(i)}}(\cdot)$ ($i = 1,2$) are weakly absolutely continuous. From $F_{X^{(i)}}(\cdot) = F_{X_+^{(i)}}(\cdot) + F_{X_-^{(i)}}(\cdot)$ ($i = 1,2$) it follows that $F_{X^{(2)}}(\cdot) = 0$, i.e. $X_{n,+}^{(2)} = 0$ ($n \in Z$). The BSVSP's $X' = X_+^{(1)}$, $X'' = X_-^{(1)}$ and $X''' = X_-^{(2)}$ have the properties described in the theorem.

Let $w \in \underline{\underline{P}}(\underline{B})$. We set

$$I_w(f) = \inf_N \inf_{f_1,\ldots,f_N} \int \sum_{k,l=0} e^{i(k-l)\mu}\, w(\mu, f_k, f_l)\, d\mu \quad (f_0 = f \in \underline{B}).$$

Let $X_0, X_{-1}, X_{-2}, \ldots$ be known and let $X_1 f$ be to be predicted. The square of the prediction error is then given by $I_{w_X}(f)$, where $w_X(\cdot, f, g)$ is the derivative of the asbolutely continuous part of $< f, F_X(\cdot)\, g >$. The BSVSP X is called to have nearly full rank if this prediction error is equal to zero only for $f = o$.

In the finite dimensional case every BSVSP X hawing nearly full rank for which $F_X(\cdot)$ is weakly absolutely continuous is regular. As Lowdenslager [8] has conjectured, this holds also in the case of a separable Hilbert space $\underline{B}$. A counterexample constructed by Douglas [3] shows that this conjecture is false, even in the case when w_X has the form (1). However, we can show :

<u>Theorem 6</u>. Let X be a BSVSP having nearly full rank and let $F_X(\cdot)$ be weakly absolutely continuous. If then measurable functions $s, t, \hat{w}_j$ ($j = 1,2,\ldots,N$) on $[-\pi, +\pi)$ and a CONS $\{e_j\}_{j=1}^N$ ($N \leq \infty$) in $\underline{B}$ exist such that

(i) $\int w_j(\mu)d\mu \leq C < \infty$

(ii) $\int (t(\mu)/s(\mu)) \ \tilde{w}_j(\mu) \ d\mu < \infty$

(iii) $\int \ln(s(\mu)/t(\mu))d\mu > -\infty$

(iv) $\int \ln t(\mu)d \mu > -\infty$

(v) $s(\cdot)\ w_X(\cdot,f,f) \leq t(\cdot)\ \tilde{w}(\cdot,f,f) \leq w_X(\cdot,f,f) \quad (f \in \underline{B})$

$$(\tilde{w}(\cdot,f,g) = \sum_{j=1}^{N} < f,e_j > \ < e_j,g > \ \tilde{w}_j(\cdot) \quad (f,g \in \underline{B}))$$

then X is regular.
The proof is based on the following lemma, which is a generalization of [9], Theorem 4.5., and the proof of which is omitted.

Lemma. Let w_1, $w_2 \in \underline{P}(\underline{B})$, and let s and t be a positive and a measurable non-negative function, respectively, such that

$$s(\cdot)\ w_2(\cdot,f,f) \leq t(\cdot)\ w_1(\cdot,f,f) \leq w_2(\cdot,f,f) \quad (f \quad \underline{B})$$

and

$$\int \ln t(\mu)\ d\mu > -\infty$$

If w_1 is factorable, then w_2 is factorable.

Proof. (of Theorem 6). From (i) it follows $\tilde{w} \in \underline{P}(\underline{B})$. Let $\tilde{\tilde{w}}_j(\cdot) = t(\cdot)/s(\cdot)\ \tilde{w}_j(\cdot)$ $(j = 1,2,\dots,N)$. For each trigonometric polynomial $q(\cdot) = \sum_{k=1}^{n} \alpha_k e^{i\cdot k}$ we have by (v)

$$\int |1+q(\mu)|^2 \tilde{\tilde{w}}_j(\mu)d\mu \geq \int |1+q(\mu)|^2 \ w_X(\mu,e_j,e_j)d\mu \geq I_{w_X}(e_j) > 0$$
$$(j = 1,2,\dots,N).$$

By (ii) and (iii), this gives $\int \ln w_j(\mu)d\mu > -\infty$ $(j = 1,\dots,N)$.
Now, by (i) there exist functions $a_j \in H^2$ such that $(1/2\pi)\ |a_j(\cdot)|^2 = \tilde{w}_j(\cdot)$ $(j = 1, \dots,N)$. This shows that the mapping $\tilde{w}$ is factorable by the operator $A^{\sim}$ defined by $(A^{\sim} f)(\cdot) :=$
$= \sum_{j=1}^{N} < f,e_j > a_j(\cdot) \quad (f \in \underline{B})$.

By the foregoing lemma, (iv) and (v) imply the factorability of w_X, i.e., the regularity of X.

Let $\underline{B}$ be a separable Hilbert space and let w be as in (1) where W is a weakly summable function on $[-\pi,+\pi]$ with values in $[\underline{B}]$ with $< f, W(\mu) f > \geq 0$ $(\mu \in [-\pi,+\pi], f \in \underline{B})$ and $W(\mu)$ is bounded invertible for almost all μ. We consider the following conditions:

(4) $\int \ln m(W(\mu))d\mu > -\infty \quad (m(W(\mu)) := \inf_{\substack{f \in \underline{B} \\ ||f||=1}} \langle f, W(\mu) f \rangle).$

(5) $\int < f, \ln W(\mu) f > d\mu \geq c \, ||f||^2 \quad (f \in \underline{B}, c > -\infty).$

(6) $\int tr \ln W(\mu) d\mu > -\infty.$

(7) $\int \ln \det W(\mu) d\mu > -\infty.$

It is well known that in the case of a finite dimensional space $\underline{B}$ these conditions are all equivalent and are necessary and sufficient for the factorability of W with $Q(0)^* Q(0)$ invertible (i.e. $I_w(f) \geq d||f||^2$ $(f \in \underline{B})$). In the infinite dimensional case only the implication (4) $\Longrightarrow$ (5) and the equivalence (6) $\Longleftrightarrow$ (7) hold (by an appropriate definition of tr and det for an operator in infinite dimensional Hilbert space). (4) is sufficient for the factorability of W with $Q(0)^* Q(0)$ invertible (Devinatz [2]), (5) is not sufficient for the factorability of W with $0 \notin \sigma_p(Q(0)^*Q(0))$ (Lax [7]). (6) and (7) are not sufficient for the factorability of W. (4) and (5) are not necessary for the factorability of W with $0 \notin \sigma_p(Q(0)^*Q(0))$, (6) and (7) are not necessary for the factorability of W with $0 \notin \sigma(Q(0)^*Q(0))$. (The corresponding counterexamples can be given with functions which values are operators of diagonal form w.r.t. a CONS.)

References

[1] J. Blum, B. Eisenberg, A note on random measures and moving averages on non-discrecte groups, Annals of Probability $\underline{1}$, 2 (1973) 336-337.

[2] A. Devinatz, The factorization of operator valued functions, Ann. of Math. $\underline{73}$ (1961) 458-495.

[3] R,G. Douglas, On factoring positive operator functions, Journ. of Math. and Mech. $\underline{16}$ (1966) 119-126.

[4] P.R. Halmos, Introduction to Hilbert space and the theory of spectral multiplicity, New York 1951.

[5] E. Hewitt, K.A. Ross, Abstract harmonic analysis I, Berlin-Göttingen-Heidelberg 1963.

[6] A.N. Kolmogorov, Stationary sequences in Hilbert space (in Russian), Bull.MGU 2, 6(1941) 1-40.

[7] P.D. Lax, On the regularity of spectral densities, Teorija verojatn. 8 (1963) 337-340.

[8] B.D. Lowdenslager, Of factoring matrix-valued functions, Ann. of Math. 78 (1963) 450-454.

[9] A.G. Miamee, H. Salehi, Factorization of positive operator valued functions on a Banach space, Indiana Math. J. 24, 2 (1974) 103-113.

[10] F. Schmidt, Verallgemeinerte stationare stochastische Prozesse auf Gruppen der Form $Z\times G^-$, Math. Nachr. 57 (1973) 337-357.

Sektion Mathematik
Technische Universität
DDR - 8027 D r e s d e n

APROPOS OF PROFESSOR MASANI'S TALK

F.H. Szafraniec

Summary. Our purpose is twofold: 1° to relate Masani's results (this Proceedings) to ours, 2° show that our results can automatically be set into the *-semigroup of actions context.

1. Let us fix some preliminaries: S is a *-semigroup with a unit, B(H) is the algebra of all bounded linear operators in a Hilbert space H, $\varphi : S \longrightarrow B(H)$ is positive definite that is

$$\sum_{i,j} (\varphi(s_i^* s_j), f_j, f_i) \geq 0$$

$s_1,\ldots,s_n \in S$, $f_1,\ldots,f_n \in H$.

In the general dilation theorem of B.Sz.-Nagy the following condition has appeared; for every $u \in S$ there is a $c(u) \geq 0$ independent of $s_1,\ldots,s_n$ in S and $f_1,\ldots,f_n$ in H such that

$$\text{(A)} \qquad \sum_{i,j} (\varphi(s_i^* u^* u s_j) f_j, f_i) \leq c(u) \sum_{i,j} (\varphi(s_i^* s_j) f_j, f_i).$$

In [6] and [8] we offered two equivalent conditions :

$$\text{(B)} \qquad ||\varphi(s)|| \leq C\,\alpha(s),$$

where α is a submultiplicative function on S ;

$$\text{(C)} \qquad \liminf_{k\to\infty} \Big(\sum_{i,j} (\varphi(s_i^*(u^* u)^{2^k} s_j) f_j, f_i)\Big)^{2^{-k}}$$

is finite and independent of $s_1,\ldots,s_n$ and $f_1,\ldots,f_n$.

The proof that (C) implies (A) was the only one thing which could meet with some difficulties. We overcame them in [6] in an elementary way making a successive use of some sort of Schwarz inequality. In [8] we isolated (B) what splits conveniently the proof of the implication (A) $\Longrightarrow$ (C) into two: (A) $\Longrightarrow$ (B) (asy, two-line proof) and (B) $\Longrightarrow$ (C) (trivial).

Masani's progress in this matter is the following condition

$$\text{(A bis)} \qquad (\varphi(s^* u^* us)f, f) \leq b(u)(\varphi(s^* s)f, f)$$

where $b(u) \geq 0$ is independent of s and f. This simplifies (A). The proof of (A bis) (see [1] and [2] ; the ealier version of [1], entitled "An explicit treatment of dilation theory" has been circulating since 1975 but it did not contain any investigations on (A) as well as (A bis)) uses the very same arguments as those of [6] though the way of proving is much more longer.

Our goal is to point out that (A bis) fits in nicely with what we have proposed. More precisely, this condition ought to be placed just between (A) and (B) , that is we have the following chain of implications: (A) $\Longrightarrow$ (A bis) $\Longrightarrow$ (B) $\Longrightarrow$(C) $\Longrightarrow$ (A). We will see in the next section how close (A bis) and (B) are.

2. To be more accurate we have to add that Masani deals rather with a *-semigroup of actions on a set than a *-semigroup itself. All what we have said so far can be carry without any change over to *-semigroup of actions. Let Z be a set and S be a *-semigroup of actions on Z. Write $u(t)$ for the action of $u \in S$ on $t \in Z$. Take a positive definite operator valued kernel K satisfying the condition

$$K(s,u(t)) = K(u^*(s),t), \quad s,t \in Z, \quad u \in S.$$

Proposition. The following conditions are equivalent:

1° $$\sum_{i,j} (K(u(t_i),u(t_j))f_j,f_i) \leq c(u) \sum_{i,j} (K(t_i,t_j)f_j,f_i)$$

where $c(u)$ is independent of $t_1,\dots,t_n$ in Z and $f_1,\dots,f_n$ in the uderlying space *);

2° $$(K(u(t),u(t))f,f) \leq b(u)(K(t,t)f,f)$$

where $b(u)$ is independent of t and f ;

3° $$||K(u(t),u(t))|| \leq C(t)\, \alpha(u)$$

where α is a submultiplicative function on S ;

4° $$\liminf_{k\to\infty} \Big(\sum_{i,j} K((u^*u)^{2^k}(t_i),(u^*u)^{2^k}(t_j))f_j,f_i)\Big)^{2^{-k-1}}$$

is finite and idependent of $t_1,\dots,t_n$ and $f_1,\dots,f_n$.

*) We do not specify whether $K(s,t)$'s act on a Hilbert space or in a Banach space as Masani's. The notation is flexible enough to allow both possibilities.

Proofs of all implications involved are just the same as those described in the preceding section. We say a couple of words about two of them. The implication $4^\circ \Longrightarrow 1^\circ$: we have the following Schwarz inequality

$$\Big|\sum_{i,j}(K(s_i,t_j)g_j,f_i)\Big|^2 \leq \sum_{i,j}(K(s_i,s_j)f_j,f_i)\sum_{i,j}(K(t_i,t_j)g_j,g_i)$$

for $s_1,\ldots,s_n,t_1,\ldots,t_n$ and $f_1,\ldots,f_n$, $g_1,\ldots,g_n$. This implies

$$\sum_{i,j}(K(u(t_i),u(t_j))f_j,f_i)^2 = \sum_{i,j}(K(t_i,u^*u(t_j))f_j,f_i)$$
$$\leq \sum_{i,j}(K(t_i,t_j)f_j,f_i)\sum_{i,j}(K(u^*u(t_i),u^*u(t_j))f_j,f_i)$$

and this leads to

$$\sum_{i,j}(K(u(t_i)u(t_j))f_j,f_i) \leq \sum_{i,j}(K(t_i,t_j)f_j,f_i)^{1-2^{-k-1}}$$
$$\times \sum_{i,j}(K((u^*u)^{2^k}(t_i),(u^*u)^{2^k}(t_j))f_j,f_i))^{2^{-k-1}}$$

Letting $n \longrightarrow \infty$ we get 1°.

The implication $2^\circ \Longrightarrow 3^\circ$: Choosing $b(u)$ to be minimal in 2° we easly check that $b(u)$ is submultiplicative. Thus we can set $\alpha(u) = b(u)$ and $C(t) = ||K(t,t)||$ to get 3°.

This shows that from 2° to 3° is not too far.

3. Examples of applications of (B) and (C) have been done in [6,7,8,9]. Here we focus on that discussed in [7]. We will try to look at it in a some-what different manner. Let S,H,φ be as in the Section 1 with one exeption : we do not require S has a unit. Suppose we are given a factorization

(F) $\qquad \varphi(s^*t) = X(s)^*\, X(t), \quad s,t \in S$

where $X(s)$ is a bounded linear map of H to another Hilbert space, say $\hat{H}$. If φ satisfies (A) then we can define an involution preserving semigroup homomorphism $T: S \longrightarrow B(H)$ such that

$$X(us) = T(u)X(s).$$

In case S has a unit this takes the form

$$X(u) = T(u)X(1)$$

and vice versa, because $T(u)$'s form a semigroup, the latter condition allows to redetermine the previous one. Viewing the second condition as of the <u>initial value</u> type we ask what will happen if S has no unit.

If S has no unit, adjoin 1, define $1^* = 1$ and denote the resulting *-semigroup by S_1. If S already has a unit, set $S_1 = S$. Then $\varphi : S \longrightarrow B(H)$ can be extended to a positive definite function on S_1 of and only if [5]

$$\text{(E)} \quad \sum_{i,j} (\varphi(s_i^* s_j)f_j,f_i) \geq c \, || \sum_i \varphi(s_i)f_i||^2 , \; \varphi(s^*) = \varphi(s)^*$$

for $s,s_1,\dots,s_n$ and $f_1,\dots,f_n$. The extended function, call it φ_1, can be defined in plenty of ways. Here we put $\varphi_1(1) = cI$, I - the identity operator in H. We have $(u \neq 1)$

$$(\varphi_1(u^*u)f,f) \leq c^{-1} \, ||\varphi(u^*u)|| \quad (cf,f) = c^{-1}||\varphi(u^*u)||(\varphi_1(1)f,f)$$

which shows that φ_1 satisfies (A bis) with $b_1(u) = \max\{b(u), c^{-1} ||\varphi(u^*u)||\}$ if so does φ with $b(u)$. Thus we come to the following

<u>Corollary.</u> Suppose φ satisfies (E) and (A) and, moreover, it factors as in (F) with $\hat{H}$ = closed linear span of $X(s)H$, $s \in S$. Then there exist :

(i) a Hilbert space $\hat{H}_1$ containing $\hat{H}$ and such that codim $\hat{H} \leq 1$;

(ii) a factorization

$$\varphi(s^*t) = X_1(s)^* X_1(t), \quad s,t \in S_1$$

where $X_1(S)$ for every $s \in S_1$ is a bounded linear map of H to $\hat{H}_1$, $X(s)$ and $X_1(s)$ are related as follows

$$X(s) = PX_1(s), \quad s \in S$$

where P is the orthogonal projection of $\hat{H}_1$ onto $\hat{H}$,

(iii) an involution preserving semigroup homomorphism $T_1 : S_1 \longrightarrow B(H_1)$ such that

$$X_1(u) = T_1(u)X_1(1), \quad u \in S_1.$$

We skip over further details calling reader's attention to the role played by the uniqueness assertion.

4. As a final remark we reflect on Masani's exploration of Steinspring-like theorems (Th.5.4 of [1], Ths. 5.5 and 6.6 of [2]). First of all we refer to Paschke's theorem [4] as to the far reaching generalization of the Stinespring dilations. Next we have to stress that both (A) and (A bis) are automatically satisfied oven in more general circumstances than those of [1,2]. Merely, the are enforced by "admissibility" of positive linear functionals on Banach star algebras (cf.[5], also [3]). Consequently, neither of these conditions prevails.

The author would like to thank Professor Masani for letting the author know both version of [1] and also for the opportunity of presenting him at the begining of October 1976 of preprints of [5,6,7,8].

References

[1] P. Masani, Dilations as propagators of Hilbertian varieties, Preprint, November 1976.

[2] -----, Propagators and dilations, this Proceedings

[3] W. Mlak, Dilations of Hilbert space operators (General theory), to appear in Dissertationes Math.

[4] W.L. Paschke, Completely positive maps on U*-algebras, Proc. Amer. Math. Soc., 34 (1972), 412-416.

[5] F.H. Szafraniec, A note on a general dilation theorem, Ann. Polon. Math. (to appear).

[6] -----, On the boundeaness condition involved in dilation theory, Bull. Acad. Polon. Sci., Sér.sci.math.,astronom. phys., 24 (1976), 877-881.

[7] -----, A general dilation theorem, ibidem, 25 (1977), 263-267.

[8] -----, Dilations on involution semigroups, Proc.Amer.Math.Soc. (to appear).

[9] -----, Boundedness in dilation theory, to appear in Banach Center Publications, the Spectral Theory volume.

Instytut Matematyki- Uniwersytet Jagielloński
ul. Reymonta 4, 30059 K r a k ó w , Poland

BOUNDEDNESS AND CONVERGENCE OF BANACH LATTICE VALUED SUBMARTINGALES

Jerzy Szulga

In this note we consider properties of Banach lattice valued submartingales and their connections with underlying structure of Banach lattice.

Definitions and problems.

In the sequel $E = (E, \|\cdot\|, \leq)$ denotes a real separable Banach lattice ; E^+ denotes the positive cone in E ; E^* is the norm dual of E. We will write, as usual, $x^+ = \sup(x,0)$, $x^- = \sup(-x,0)$, $|x| = \sup(x,-x)$. Let $L = L(E)$ denote the Banach lattice of all strongly measurable and integrable E-valued functions defined on some probability space $(\Omega, \mathfrak{F}, P)$, with the usual L-norm, $\|X\|_L = E\|X\|$. Let $\{\mathfrak{F}_n\}$ be an increasing family of sub-σ-fields of $\mathfrak{F}$. A sequence $\{X_n\} \subset L(E)$ is said to be a <u>submartingale</u> if for each n, X_n is $\mathfrak{F}_n$-measurable and $E(X_{n+1} \mid \mathfrak{F}_n) \geq X_n$. Clearly, $\{X_n\}$ is a martingale if both $\{X_n\}$ and $\{-X_n\}$ are submartingales.
The notion of a lattice valied submartingale is due to F.Scalora. All the definitions and facts concerning Banach lattice can be found in a H.H. Schaefer's monography [5].

What we are interested in is how the Doob fundamental submartingale theorem (cf. [3], p.63) carries over to the case of Banach lattice valued objects. It is quite easy to check that, in general, the <u>Doob condition</u> for submartingale $\{X_n\}$:

$$\sup_n E\|X_n^+\| < \infty$$

doesn't suffice to a.s. convergence or even L-boundedness of $\{X_n\}$([66], Ex. 3.1). On the other hand under some stronger conditions concerning a submartingale one can obtain an analogue of Chaterji Martingale Theorem ([1], [6]).

Now, our main interest is to give a characterization of a class of Banach lattices which satisfy the Doob Theorem.

That is, we say that E satisfies a <u>submartingale boundedness theorem</u> (resp. convergence theorem) and we will write $E \in SBT$ (resp. $E \in SCT$) if each E-valued submartingale satisfying the Doob condition is bounded in $L(E)$ (resp. converges a.s. to some integrable random vector).
The following class of Banach lattices will be useful : E is said to be an <u>AL-space,</u> if its norm is additive on E^+ :

$$||x + y|| = ||x|| + ||y|| \quad \text{for all } x,y \geq E^+ .$$

Now we have

<u>Theorem</u> 1. The following conditions are equivalent :

(1) $E \in SBT$;

(2) Any positive increasing sequence $\{A_n\} \subset L(E)$ is L-bounded provided $\sup_n ||E\, A_n|| < \infty$,

(2') The as in (2) but additionaly $\{A_n\}$ is of the form :

$$A_n = \sum_{i=1}^{n} f_i\, x_i \ ;$$

where $\{x_n\} \subset E^+$ and $\{f_n\}$ is a sequence of real positive independent random variables ;

(3) E is isomorphic (as a Banach lattice) to an AL-space.

<u>Proof</u> : We will show
$$\begin{array}{ccccc} & \nearrow & (2') & \nwarrow & \\ (1) & & \Downarrow & & (2). \\ & \nwarrow & (3) & \nearrow & \end{array}$$

(1) ==> (2'). If $\{x_n\}$ and $\{f_n\}$ are as required then a martingale $\{X_n\}$ defined as a sum

$$X_n = \sum_{i=1}^{n} (E\, f_i - f_i)x_i \ ,$$

satisfies the Doob condition ; hence (1) ==> (2') holds

(2) ==> (2') is evident.

If $||.||$ denotes the AL-norm, the

(i) $E\, ||X|| = ||\, E|X|\, ||$

for all $X \in L(E)$ ((i) holds true for the set of step functions, which is dense in $L(E)$).

Hence (3) ==> (2) and (3) ==> (1) hold

(2') ==> (3). For $\underline{x} = \{x_n\} \in E^N$ put

$$||\underline{x}||_1 = \sup_n || \sum_{i=1}^{n} E\, f_i\, |x_i| \;||\,,$$

$$||\underline{x}||_2 = \sup_n || \sum_{i=1}^{n} f_i\, |x_i| \;||,$$

Then $H_i \overset{df}{=} \{ \underline{x} \in E^N : ||\underline{x}||_i < \infty\}$, $i = 1,2$, is the Banach lattice under the canonical ordering. By the assumptions, $H_1 \subset H_2$ and by the Closed Graph Theorem we have

(ii) $||\underline{x}||_2 \leq C||\underline{x}||_1$ for all $\underline{x} \in H_1$.

Now let $\{f_n^o\}$ be a sequence of real independent random variables such that $P(f_n^o = 0) = 1 - p_n$, $P(f_n^o = \frac{1}{p_n}) = p_n$ where $\{p_n\}$ is a sequence of reals s.t. $0 < p_n < 1$ and $\prod_{i=1}^{\infty}(1-p_i) = p > 0$.
Then $\{f_n^o\}$ has the property :

(iii) $\sum_{i=1}^{n} ||x_i|| \leq M\, E\, || \sum_{i=1}^{n} f_i^o\, x_i ||$ for all $x_1,\ldots,x_n \in E$

Now by (ii) and (iii) we have

(iv) $\sum_{i=1}^{n} ||x_i|| \leq C\, M\, || \sum_{i=1}^{n} x_i||$ for all $x_1,\ldots,x_n \in E^+$

Condition (iv) is the Schlotterbeck characterization of B. lattices isomorphic to some AL-space (cf. [5], p.242).

That is, the norm

$$||x||_* = \sup\{\sum_{i=1}^{n} ||x_i|| : x_1,\ldots,x_n \in E^+ \text{ and } \sum_{i=1}^{n} x_i \leq |x|\,\}$$

is monotone, complete, equivalent to the original one and additive on E^+. Q.E.D.

<u>Remark</u> : The above equivalences remain true, if we replace (2) by a priori weaker condition :

(2a) $\{A_n\} \subset L(E)$ (positive, increasing) is bounded in $L(E)$ if the set $\{E\, A_n\}$ E is lattice bounded (i.e. $E\, A_n \leq x_0 \in E$)

Indeed, clearly (2a) is weaker than (2). If E has the property (2a), then E is so called KB-space (no sublattice of E is isomorphic to c_0). To see that, it suffices to construct a sequence $\{A_n\}$ of c_0-valued positive random vectors such that A_n increases,

$$E\, A_n \leq x_0 \in c_0 \quad \text{and} \quad \sup_n E\, ||A_n||_{c_0} = \infty .$$

If we put

$$A_n = \sum_{i=1}^{n} \frac{1}{i\, a_i} I_{B_i}\, e_i ,$$

where $\{e_i\}$ is the standart basis of c_0, $\{B_i\}$ is a sequence of disjoint events with $P(B_i) = a_i > 0$ and I_B is the indicator function of the set $B \subset \Omega$; then $\{A_n\}$ is as required.
Now in KB-spaces every positive increasing norm bounded sequence is lattice bounded (cf [5], Th. 5.9).

Corollary. $E \in SCT$ iff $E \in SBT$ and E has Radon-Nikodym property. Hence _$E \in SCT$ iff E is isomorphic to a sublattice of l_1_.

Indeed, the first statement follows from the Chaterji Martingale Theorem. The second one :
If $E \in SCT$, then by Theorem 1, E is isomorphic to an AL-space, thus by Kakutani representation theorem ([2]) E is isomorphic to $L_1(S,\Sigma,\mu)$ the Banach lattice of all real integrable functions on some measure space (S,Σ,μ). On the other hand $L_1(S,\Sigma,\mu)$ has no Radon-Nikodym property unless μ is purely atomic. That yields the desired result.

Conversely, if $E = l_1$ and $\{X_n\}$ is an l_1-valued submartingale satisfying the Doob condition, then we can decompose $\{X_n\}$ in the form

$$X_n = M_n + A_n ,$$

where $\{M_n\}$ is a martingale and $\{A_n\}$ is a positive increasing sequence. Now $\{M_n\}$ also satisfies the Doob condition, hence by (1) (see the Proof of Th. 1.) and the martingale property $\{M_n\}$ is $L(l_1)$ -bounded, hence converges a.s.

Then $\{A_n\}$ is also $L(l_1)$ - bounded, hence $\{A_n(w)\}$ is l_1-bounded for almost all $w \in \Omega$, therefore it converges a.s., since l_1 is KB-space. This completes the proof.

Cone absolutely summing operators.

The introduced technique gives us possibility of characterization of other "lattice objects".

Let F be a Banach space. A linear continuous operator $T : E \longrightarrow F$ is said to be _cone absolutely summing_, if the image by T of each positive summable sequence in E, is absolutely summable in F.

<u>Theorem 2.</u> A continuous operator $T : E \longrightarrow F$ is c.a.s. if any of the following conditions is satisfied :

(1) For any E-valued submartingale $\{X_n\}$, the image $\{T X_n\}$ is L(F) -bounded ;

(2) For any positive increasing sequence $\{A_n\} \subset L(E)$, $\{T A_n\}$ converges a.s. and in L(F).

<u>Proof.</u> Let T be c.a.s.

ad (1). Let submartingale $\{X_n\}$ satisfy the Doob condition. Since T is c.a.s., then there exists $x^* \in E^*$, such that for all $x \in E$

$$||T x|| \leq < x^*, |x| >$$

(cf.[5], p.244). Hence, since $|x| = 2 x^+ - x$,

$$\sup_n E||T X_n|| \leq 2 \sup_n E ||X_n^+|| + E ||X_1|| < \infty$$

ad (2). Let $A_n \geq 0$ be increasing. Since T is c.a.s., then there exists an AL-space G and continuous linear operators $T_1 : E \longrightarrow G$, $T_1 \geq 0$ and $T_2 : G \longrightarrow F$ such that $T = T_2 T_1$. The sequence $\{T_1 A_n\} \subset L(G)$ is positive, increasing and bounded in L(G) (by (i), the proof of Th. 1.)
Since every AL-space is a KB-space, $\{T_1 A_n\}$ converges a.s. and hence by the Lebesgue Theorem it converges also in L(G). Since T_2 is continuous (2) follows.
Let now T satisfy (1) or (2). Let us define a positive increasing sequence

$$A_n = \sum_{i=1}^{n} f_i^o x_i$$

or a martingale satisfying the Doob condition

$$M_n = \sum_{i=1}^{n} x_i - A_n ,$$

where $\{x_n\} \subset E^+$ is a summable sequence and $\{f_n^o\}$ is defined as in the proof of Th.1. Any of the conditions (1), (2) yields

$$\sup_n E || T \sum_{i=1}^{n} f_i^o x_i || < \infty$$

Therefore by (iii)

$$\sum_{i=1}^{\infty} ||T x_i|| \leq M \sup_n E ||T \sum_{i=1}^{n} f_i^o x_i|| < \infty$$

Q.E.D.

Remark : The class of c.a.s. operators can be also characterized in terms of positive martingales, supermartingales, etc., however arguments are similar and differences are not essential.

References

[1] S.D. Chaterji, Martingale convergence and the Radon-Nikodym theorem in Banach spaces, Math. Scand., 22 (1968), 21-41.

[2] S. Kakutani, Concrete representation of abstract L-spaces and the mean ergodic theorem, Ann. of Math., 42 (1941), 523-537.

[3] J. Neveu, Martingales a Temps Discrete, Paris 1972.

[4] F. Scalora, Abstract martingale convergence theorem, Pacific. J. Math., 11 (1961), 347-374.

[5] H.H. Schaefer, Banach Lattices and Positive Operators, Berlin-Heidelberg-New York 1974.

[6] J. Szulga - W.A.Woyczyński, Convergence of submartingales in Banach lattices, The Ann. of Prob., 4, No.3, (1976), 464-469.

Mathematical Institute
Wrocław University
Pl.Grunwaldzki 2/4
50-384 Wrocław
P o l a n d

MULTIPLICITY THEORY FOR RANDOM FIELDS USING QUANTUM MECHANICAL METHODS

Dag Tjøstheim

1. Introduction

For nonstationary random processes extensive representation and multiplicity theoretic results exist in the general multivariate case. We refer to papers by Cramér [4] and Hida [9], Kallianpur and Mandrekar [10] and Tjøstheim [19],[20]. In comparison, the representation and multiplicity theory of random fields is rather incomplete. Although representations of nonhomogeneous random fields has been considered by Rao [15], his starting point is somewhat different from that of Cramér and Hida. To our knowledge, results analogous to those of [4] and [9] have not been established for p.n.d. random fields. Univariate homogeneous generalized fields are covered in a paper by Urbanik [22]. Similar results do not seem to exist for the multivariate case.

It will be our purpose to try to fill these gaps in the theory. We will consider multivariate (possibly infinitedimensional) random fields, both ordinary and generalized. The representation theory will be based on a decomposition in a direct integral of Hilbert spaces induced by a commuting family of self-adjoint operators. Such methods are familiar in quantum mechanics (see for example [18] and references therein). Such techniques were introduced in the context of random processes/fields (in spectral domain) in Tjøstheim [21]. The direct integral decomposition associated with a self-adjoint operator was established by von Neumann [13] and may be viewed as an alternative approach to the Hellinger-Hahn theory of spectral multiplicity. The latter theory was the basis for the results obtained in [4] and [9].
As will be seen in Section 4 in the special case of a homogeneous field, the quantum mechanical concept of a Schrødinger system of position-momentum operators plays an important role in the second order theory. The use of quantum mechanical commutation relations in the multiplicity theory of general multivariate wide sense stationary processes was initiated in Tjøstheim [19], [20].

See also Gustafson and Misra [7]. For ealier papers in the multiplicity/time domain representation theory of wide sense stationary processes we refer to Hanner [8], Masani and Robertson [11] for the univariate case and to Kallianpur and Mandrekar [10], Robertson [17] for the multivariate case.

2. Preliminaries and Notation

We denote by H the Hilbert space of all complex-valued random variables having a finite second moment, and where the inner product is defined by $(F,G) = E\{F\,\overline{G}\}$, F and $G \in H$. Let K be the space of complex-valued test functions on R^n having a compact support. It is assumed that K is equipped with the usual Schwartz topology [24, p. 28]. Let L be a parameter set. We shall say that $F_\lambda(\Phi)$, $\lambda \in L$, $\Phi \in K$, is a second order multivariate generalized random field over R^n if for each λ and Φ, $F_\lambda(\Phi) \in H$ and the mapping $\Phi \to F_\lambda(\Phi)$, $\lambda \in L$, of K into H is linear and continuous. If $L = R^m$ and $F_\lambda(\Phi)$ is linear in λ for each $\Phi \in K$, $F_\lambda(\Phi)$ is a m-dimensional generalized field. The field $F_\lambda(\Phi)$ will be said to be homogeneous if for arbitrary $\lambda, \mu \in L$; $\Phi, \Psi \in K$ and $y \in R^n$

$$E\{F_\lambda(\tau(y)\Phi)\ \overline{F_\mu(\tau(y)\Psi))}\} = E\{F_\lambda(\Phi)\ \overline{F_\mu(\Psi)}\}$$

where $\quad \tau(y)\Phi(x) = \Phi(x+y), \quad x \in R^n.$

Denote by $H(F) \subset H$ the Hilbert space generated by $F_\lambda(\Phi)$ as λ and Φ run through L and K respectively. In all of the following it will be assumed that $H(F)$ is separable. Sufficient conditions on L and $F_\lambda(\Phi)$ for this to be true are: L is a Hausdorff space satisfying the second countability axiom and $F_\lambda(\Phi)$ is continuous in quadratic mean (q.m.) relative to the topology of L for each $\Phi \in K$. The separability follows by some minor adjustments of the arguments used in the proof of Lemma 2.1 of [10].

Let $S_k(t)$, $k = 1,\dots,n$; $t \in (-\infty, \infty)$, be the half-space given by

$$S_k(t) = \{y = (y_1,\dots,y_n) \in R^n \ : \ y_k \leq t\}$$

and define $S_-(x) = \bigcap_{k=1}^{n} S_k(x_k)$ where $x = (x_1,\dots,x_n) \in R^n$.
Denote by $H_k(F,x_k)$ and $H_-(F,x)$ the subspaces of $H(F)$ generated by all elements $F_\lambda(\Phi)$ with the constraints $\mathrm{supp}(\Phi) \subset S_k(x_k)$ and $\mathrm{supp}(\Phi) \subset S_-(x)$ respectively. Let $H_k(F,t+) = \bigcap_{s>t} H_k(F,s)$ and let E_t^k be the projection operator on $H_k(F,t+)$ for $t \in (-\infty,\infty)$. (We note that $H_k(F,t+) = H_k(F,t)$ for a homogeneous field.)
If $\bigcap_{t\in(-\infty,\infty)} H_k(F,t+) = 0$, the chain of spaces $H_k(F,t+)$, $t \in (-\infty,\infty)$, defines a self-adjoint operator Q_k having E_t^k, $t \in (-\infty,\infty)$, as its resolution of identity. The operator Q_k, $1 \leq k \leq n$, will be called the <u>kth coordinate operator</u> of the field. For the random processes case $1 = k = n$, and following [20] we use the notation T instead of Q and speak about the time operator T of the process.

We use the notation $H_t^k = H_k(F,t+)$. Let $s \leq t$ and let $H_k(s,t] = H_t^k \cap H_s^{k\perp}$ where the symbol $\perp$ denotes the operation of taking orthogonal complements. Similarly we define $H_k(-\infty,t] = H_t^k \cap H_{-\infty}^{k\perp}$, $H_k(s,\infty) = H_s^{k\perp}$, where $H_{-\infty}^k = \bigcap_{s\in(-\infty,\infty)} H_s^k$. Let $\Delta = (s_1,t_1]\times\dots\times(s_n,t_n]$ where s_k may be $-\infty$ and t_k $+\infty$. The space $H_{in}(\Delta) = \bigcap_{k=1}^{n} H_k(s_k,t_k]$ may be interpreted as the space spanned by the innovations received by the field in the set $\Delta \subseteq R^n$. Following [21] the field $F_\lambda(\Phi)$ will be said to be purely non-deterministic (p.n.d.) if $H(F)$ is spanned by the totality of innovations received by the field. More precisely, let I be an index set and

$$\Delta^\alpha = (s_1^\alpha, t_1^\alpha] \times \dots \times (s_n^\alpha, t_n^\alpha], \qquad \alpha \in I$$

Then $F_\lambda(\Phi)$ is said to be p.n.d. if for every collection of sets Δ^α, $\alpha \in I$, for which $\bigcup_{\alpha\in I} \Delta^\alpha = R^n$, we have that the Hilbert space generated by $\bigcup_{\alpha\in I} H_{in}(\Delta^\alpha)$ equals $H(F)$. If $F_\lambda(\Phi)$ is p.n.d., then $H_{in}(R^n) = \bigcap_{k=1}^{n} H_k(-\infty,\infty) = H(F)$. Thus for each k, $H_{-\infty}^{k\perp} = H(F)$, or $H_{-\infty}^k = \bigcap_{s\in(-\infty,\infty)} H_s^k = \{0\}$, and the coordinate operators Q_k are well defined.

For $n = 1$ the conditions a) : $\bigcup_{\alpha \in I} \Delta^{\alpha} = R^n \Longrightarrow$ the Hilbert space generated by $\bigcup_{\alpha \in I} H_{in}(\Delta^{\alpha})$ equals $H(F)$ and b): $H^k_{-\infty} = \{0\}$, $k = 1,\ldots,n$ are equivalent. Thus the above definition reduces to the familiar definition of p.n.d. in the random process case. For $n > 1$ b) does not necessarily imply a).

Let $\Delta_i = (s^i_1, t^i_1] \times \ldots \times (s^i_n, t^i_n]$, $i = 1,2$ be disjoint. Then there exists a j, $1 \leq j \leq n$, such that $(s^1_j, t^1_j]$ and $(s^2_j, t^2_j]$ are disjoint and consequently $H_j(s^1_j, t^1_j] \perp H_j(\,^2_j, t^2_j]$. But $H(\Delta_i) \subset H_j(s^i_j, t^i_j]$, $i = 1,2$ and thus $H(\Delta_1) \perp H(\Delta_2)$. It is now easy to prove that the p.n.d. property implies that the coordinate operators form a commuting family. Indeed, let $1 \leq i, \; j \leq n$, $i \neq j$, and let s and t be real. We define

$$\Delta_1 = \{x = (x_1,\ldots,x_n) \in R^n : \; x_i \leq s, \; x_j \leq t\}$$

$$\Delta_2 = \{x = (x_1,\ldots,x_n) \in R^n : \; x_i \leq s, \; x_j > t\}$$

$$\Delta_3 = \{x = (x_1,\ldots,x_n) \in R^n : \; x_i > s, \; x_j \leq t\}$$

$$\Delta_4 = \{x = (x_1,\ldots,x_n) \in R^n : \; x_i > s, \; x_j > t\}$$

Then $R^n = \bigcup_{i=1}^{4} \Delta_i$, and from the p.n.d. property and the above orthogonality property it follows that $H(F) = \sum_{j=1}^{4} \oplus H(\Delta_j)$.

Let $F \in H(\Delta_1)$. Since F_λ is p.n.d., $H(\Delta_1) = H^i_s \cap H^j_t$. Thus $E^i_s E^j_t F = E^j_t E^i_s F = F$. Let $F \in H(\Delta_2) = H^i_s \cap H^{j\perp}_t$. Then $E^i_s E^j_t F = 0$ and $E^j_t E^i_s F = E^j_t F = 0$. Similarly for $F \in H(\Delta_3)$ and $F \in H(\Delta_4)$ we have $E^i_s \; E^j_t \; F = E^j_t \; E^i_s \; F = 0$. Since $H(F) = \sum_{j=1}^{4} \oplus H(\Delta_j)$ it follows that $E^i_s \, E^j_t = E^j_t \, E^i_s$ on $H(F)$. The coordinate operators Q_i and Q_j therefore commute and $\{Q_1,\ldots,Q_n\}$ form a commuting family in $H(F)$.

We shall say that $F_\lambda(x)$, $\lambda \in L$, $x \in R^n$, is a second order multivariate ordinary random field if for each λ and x, $F_\lambda(x) \in H$. We define $H(F) \subset H$ as the closure in H of the linear hull of the set of elements $F_\lambda(x)$, $\lambda \in L$, $x \in R^n$. It will always be assumed that $H(F)$ is separable. The following conditions imply separability : i) L is a Hausdorff space satysfying the second countability axiom, and $F_\lambda(x)$ is continuous in q.m. relative to the topology of L for each $x \in R^n$. ii) $F_\lambda(x)$ has a countable number of q.m. discontinuities. For a random process $Y_\lambda(t)$, $t \in (-\infty,\infty)$, condition ii) can [4] be replaced by the requirement that the limits $Y_\lambda(t-)$ and $Y_\lambda(t+)$, $t \in (-\infty,\infty)$, exist in q.m. We say that the field $F_\lambda(x)$ is homogeneous if for arbitrary x,y and $z \in R^n$,

$$E\{F_\lambda(x+z)\ \overline{F_\lambda(y+z)}\} = E\{F_\lambda(x)\ \overline{F_\lambda(y)}\}$$

We denote by $H_-(F,x)$ the subspace of $H(F)$ generated by the set of elements $F_\lambda(y)$, $\lambda \in L$, $y \in S_-(x)$. The concept of a p.n.d. field and the coordinate operators of a p.n.d. field $F_\lambda(x)$ are now defined as above. Finally, we note, Yaglom [23], that with an ordinary field $F_\lambda(x)$ continuous in q.m. for each $\lambda \in L$, we can associate a corresponding generalized field $F_\lambda(\Phi)$.

3. Multiplicity theory of Random Fields

The commutativity of the coordinate operators of a p.n.d. field can be used [13], [21 , proof of Theorem 2.1] to obtain a realization $H(F) \leftrightarrow \hat{H} = \int \hat{H}(x)\, d\mu(x)$ of $H(F)$ as a direct integral of Hilbert spaces. Denote by $d(x)$ the dimension of $\hat{H}(x)$. Then [13] the function $d: x \longrightarrow d(x)$ is μ-measurable. Let $A_i \subset R^n$, $i = 0,1,\ldots$ be the sets such that $d(x) = i$ for $x \in A_i$. Then $R^n = \bigcup_{i=0}^{\infty} A_i$. Denote A' the set obtained from $\bigcup_{i=0}^{\infty} A_i$ by deleting those A_i for which $\mu(A_i) = 0$. The <u>coordinate multiplicity</u> of the random field $F_\lambda(\Phi)$ is defined as $d_Q = \sup_{x \in A'} d(x)$. From the uniqueness properties [13], [21] of the isomorphism V it follows that d_Q does not depend on the particular realization $H(F) \leftrightarrow \hat{H}$ as a direct integral, and the concept of coordinate multiplicity is therefore well-defined. The field $F_\lambda(\Phi)$ is said to have uniform coordinate multiplicity if μ is equivalent to Lebesgue measure on R^n and $d(x)$ = constant for μ-measure a.e. In the following theorem we obtain a decomposition corresponding to the Cramér-Hida [4], [9] innovations representation of a p.n.d. random process. The proof is essentially identical to that of Theorem 3.2 of [21] and will therefore only be sketched.

Theorem 3.1.: Let $F_\lambda(\Phi)$, $\lambda \in L$, $\Phi \in K$ be a p.n.d generalized field. For each $\Phi \in K$ let $y(\Phi) = \{y_1(\Phi),\dots,y_n(\Phi)\}$ where $y_k(\Phi) = \{\sup x_k : \Phi(x_1,\dots,x_k,\dots,x_n) \neq 0$, when x_j, $= 1,\dots,n$; $j \neq k$, vary over $(-\infty,\infty)\}$. Then for each λ and Φ there exists a decomposition such that with probability one

$$F_\lambda(\Phi) = \sum_{i=1}^{d_Q} \int_{S_-(y(\Phi))} g_\lambda^i(\Phi;x)\; dZ_i(x) \qquad (1)$$

where d_Q is the coordinate multiplicity of $F_\lambda(\Phi)$, and where $Z_i(\Delta)$, $i = 1,\dots,d_Q$, Δ a Borel set, are random measures over the Borel sets of R^n having the properties

i) $E\{Z_j(\Delta_1)\, \overline{Z_k(\Delta_2)}\} = \delta_{jk}\; E|Z_k(\Delta_1 \cap \Delta_2)|^2 < \infty$

for arbitrary Borel sets Δ_1 and Δ_2 of R^n.

ii) $H_-(F,y+) = \sum_{i=1}^{d_Q} \oplus H_-(Z_i,y)$

where $H_-(Z_i,y) \subset H(F)$ is the Hilbert space generated by the linear hull of the set of elements $Z_i(\Delta)$, where Δ runs through all Borel sets such that $\Delta \subset S_-(y)$.

($H_-(F,y+)$ is the Hilbert space generated by $\bigcap_{k=1}^{n} H(F,y_k+)$)

Proof: Let $H(F) \leftrightarrow \hat{H} = \int_{R^n} \hat{H}(x)\, d\mu(x)$ be a direct integral decomposition of $H(F)$ induced by the set of coordinate operators $\{Q_1,\dots,Q_n\}$ of $F_\lambda(\Phi)$. Let $\hat{Z}_i(x)$, $i = 1,\dots,d(x)$ be an orthonormal basis in $\hat{H}(x)$, and let $\hat{Z}_i$, $i = 1,2,\dots$, be the μ-measurable vector fields defined by $\hat{Z}_i : x \longrightarrow \hat{Z}_i(x)$, where we put $\hat{Z}_i(x) = 0$ for $i > d(x)$. Then

$$||\hat{Z}_i||^2 = \int_{R^n} ||\hat{Z}_i(x)||^2_x \, d\mu(x) \le \mu(R^n) < \infty$$

where the finiteness of $\mu(R^n)$ follows from [13]. Hence $\hat{Z}_i \in \hat{H}$ for $1, 2, \ldots$ Let d_Q be the coordinate multiplicity of $F_\lambda(\Phi)$. Clearly $||\hat{Z}_i||^2 = 0$ for $i > d_Q$. Denote by Z_i, $i = 1, \ldots, d_Q$ the random variables corresponding to $\hat{Z}_i$ using the identification $H(F) \longleftrightarrow \hat{H}$, and let E^k_t, $t \in (-\infty, \infty)$, be the resolution of identi associated with Q_k. Consider the random rectangle function

$$Z_i(\Delta \, x \ldots x \Delta_n) = \prod_{k=1}^n E^k(\Delta_k) \, Z_i \quad \text{where } \Delta_k, \quad k = 1, \ldots, n \quad \text{are}$$

Borel sets in $(-\infty, \infty)$. It is not difficult to show that this set function can be extended to a random measure $Z_i(\Delta)$ over the Borel sets Δ of R^n. Clearly

$$Z_i(\Delta) \longleftrightarrow \{\chi_\Delta(x) \, \hat{Z}_i(x)\} \tag{2}$$

where $\chi_\Delta(x) = 1$ for $x \in \Delta$ and zero otherwise.

The representation (1) is now proved by decomposing $F_\lambda(\Phi) \longleftrightarrow$ $\hat{F}_\lambda(\Phi) = \{\hat{F}_\lambda(\Phi; x)\}$ along the basis $\hat{Z}_i(x)$ of $\hat{H}(x)$. Thus $g^i_\lambda(\Phi; x) = (\hat{F}_\lambda(\Phi; x), \hat{Z}_i(x))_x$ where $(,)_x$ is the inner product in $\hat{H}(x)$. Property i) of the theorem follows directly from the definition of $Z_i(\Delta)$. Denote by $\hat{H}_y$, $y = (y_1, \ldots, y_n) \in R^n$, the subspace of $\hat{H}$ spanned by those $\hat{h} = \{\hat{h}(x)\} \in \hat{H}$ for which $\hat{h}(x) = 0$ for $x \in \overline{S_-(y)}$, where $\overline{S_-(y)}$ is the complement of $S_-(y)$. Similarly, denote by $\chi_{S(y_k)}$ the characteristic function of $S(y_k)$. Since the projector $E^k_{y_k}$ in $H(F)$ corresponds to multiplication by $\chi_{S(y_k)}(x)$ in $\hat{H}(x)$, and since $E^k_{y_k} H(F) = H(F, y_k+)$, it is not difficult to verify that $H_-(F, y+) \longleftrightarrow \hat{H}_y$. From the relation (2) we have that

$$\sum_{i=1}^{d_Q} \oplus H_-(Z_i, y) \longleftrightarrow \hat{H}_y \quad \text{and property}$$

ii) of the theorem immediately follows. ||

We note from the uniqueness part of the isomorphism $H(F) \longleftrightarrow$ $\int \hat{H}(x) \, d\mu(x)$ that the representation (1) as constructed in the preceding proof is uniquely determined up to unitary equivalence.

Consider a p.n.d. random process $Y_\lambda(\Phi)$ and denote by $T = Q$ the time operator of the process. The multiplicity of $Y_\lambda(\Phi)$ is defined [20] as the spectral multiplicity of T in $H(Y)$. Here the spectral multiplicity of T is defined [1, p. 204] as the minimal dimension of its generating subspaces: Let E_t, $t \in (-\infty,\infty)$, be the resolution of identity of T. A subspace G of $H(Y)$ will be called a generating subspace of T if the linear hull of the set of elements $E(\Delta)\,G$, where Δ runs through all Borel sets $(-\infty,\infty)$ is dense in $H(Y)$. If no finite dimensional generating subspace can be found, the spectral multiplicity of T is said to be infinite. For the univariate case Chi [3] established a representation of $Y_\lambda(\Phi)$ similar to the one in Eq. (1), where the multiplicity of $Y_\lambda(\Phi)$ corresponded to d_Q of Eq.(1). We now show that for a random process $Y_\lambda(\Phi)$ the coordinate multiplicity (or "time multiplicity") is identical to the multiplicity of $Y_\lambda(\Phi)$. We thus obtain as a special case of Theorem 3.1, a multivariate extension of Theorem 2.1 in [3].

__Theorem 3.2:__ Let $Y_\lambda(\Phi)$, $\lambda \in L$, $\Phi \in K$, be a p.n.d. random process. Then the coordinate multiplicity d_Q appearing in Eq. (1) is identical to the multiplicity of $Y_\lambda(\Phi)$.

__Proof:__ Let $H(Y) \overset{V}{\longleftrightarrow} \hat{H} = \int_{-\infty}^{\infty} \hat{H}(s)\, d\mu(s)$ be a direct integral decomposition induced by the self-adjoint operator T. Denote by d_Q and M the coordinate multiplicity and multiplicity of $Y_\lambda(\Phi)$ respectively. Let G be the space spanned by the variables $Z_1,\dots,Z_{d_Q}$ where we have used the notation of the proof of Theorem 3.1. Using the representation of operator functions of type $v(T)$ in $_{d_Q}\hat{H}$ it results that every $Y \in H(Y)$ can be expressed as $Y = \sum_{i=1}^{d_Q} v_i(T)\ Z_i$ where $v_i(s) = (Y, Z_i(s))_s$.

It follows that G is a generating subspace of T and $M \leq d_Q$. If M is infinite the proof is complete. Assume M finite and $M < d_Q$, and let G be a generating subspace of T of dimension M. Let $g_1,\dots,g_M$ be a basis in G, and let $g_i \overset{V}{\longleftrightarrow} \hat{g}_i = \{\hat{g}_i(s)\}$. Denote by $H'(s)$ the subspace of $\hat{H}(s)$ generated by $\hat{g}_i(s)$, $i = 1,\dots,M$. Since $M < d_Q$ there exists an element $\hat{Y} = \{\hat{Y}(s)\} \neq 0$

in $\hat{H}$ such that $(\hat{g}_i(s), \hat{Y}(s))_s = 0$ (choose $\hat{Y}(s) = 0$ for those s for which $H'(s) = \hat{H}(s)$). However, if E_s, $s \in (-\infty,\infty)$, is the resolution of identity associated with T, then $E(\Delta)\, g_i \leftrightarrow \{\chi_\Delta(s)\, \hat{g}_i(s)\}$, and the random variable Y in $H(F)$ defined by $Y \leftrightarrow \hat{Y}$ is therefore orthogonal to the linear hull of the set of elements $E(\Delta)\, g_i$, $i = 1,\dots,M$; Δ a Borel set. This contradicts the fact that M is the spectral multiplicity of T.||

From Theorem 3.1 it follows that $H\_(F,y) \subset \sum_{i=1}^{d_Q} \oplus H\_(Z_i,y)$. For some purposes it may be desirable to have a representation

$$F_\lambda(\Phi) = \sum_{i=1}^{d_Q} \int_{S\_(y(\Phi))} \tilde{g}^i_\lambda(\Phi;x)\, d\tilde{Z}_i(x) \qquad (3)$$

where the random measures $\tilde{Z}_i(\Delta)$, $i = 1,\dots,d_Q$ have the some properties as in Theorem 3.1 except that ii) is replaced by

$$\text{ii}')\quad H\_(F,y) = \sum_{i=1}^{d_Q} \oplus H\_(\tilde{Z}_i,y)$$

Given a representation (1) satisfying ii) it is not difficult (applying the technique of the proof of Theorem 3.3 of [10]) to obtain a representation (3) satisfying ii'). In the random process case the representations (1) and (3) are usually termed canonical and proper canonical respectively.

Let $\Psi \in K$ be arbitrary. The optimal (in least squares sense) linear predictor of $F_\lambda(\Psi)$, given $F_\lambda(\Phi)$ for all Φ with supp $\Phi \subset S\_(y)$ can then be represented as

$$F^0_\lambda(\Psi) = \sum_{i=1}^{d_Q} \int_{S\_(y)} \tilde{g}^i(\Psi;x)\, d\tilde{Z}_i(x)$$

and the prediction error is

$$E|F^0_\lambda(\Psi) - F_\lambda(\Psi)|^2 = \sum_{i=1}^{d_Q} \int_{S-(y(\Psi))\cap \overline{S\_(y)}} |g^i_\lambda(\Psi;x)|^2\, d\tilde{\rho}_i(x)$$

where $\tilde{\rho}_i(\Delta) = E|\tilde{Z}_i(\Delta)|^2$, Δ a Borel set in R^n.

A representation similar to Eq. (1) can be established also for an ordinary random field $F_\lambda(x)$, $\lambda \in L$, $x \in R^n$. The proof is essentially as before, and we content ourselves by stating the result.

<u>Theorem 3.1'</u>: Let $F_\lambda(y)$, $\lambda \in L$, $y \in R^n$ be a p.n.d. (ordinary) random field. Then for each λ and y there exists a decomposition such that with probability one

$$F_\lambda(y) = \sum_{i=1}^{d_Q} \int_{S_-(y)} g_\lambda^i(y,x)\, dZ_i(x) \qquad (1')$$

where d_Q is the coordinate multiplicity of $F_\lambda(y)$ and $Z_i(\Delta)$, $i = 1,\dots,d_Q$; Δ a Bore set in R^n, are random measures having the properties i) and ii) of Theorem 3.1.

Theorem 3.1' generalizes the results obtained by Cramér [4] and Hida [9] for an ordinary p.n.d. random process. A proper canonical representation

$$F_\lambda(y) = \sum_{i=1}^{d_Q} \int_{S_-(y)} \tilde{g}_\lambda^i\,(y,\ )\, d\tilde{Z}_i(x) \qquad (3')$$

with $\tilde{Z}_i(\Delta)$, $i = 1,\dots,d_Q$ satisfying ii') can be established as before.

Let $Y_1(t),\dots,Y_p(t)$, $t \in (-\infty,\infty)$, be univariate p.n.d. random processes of multiplicities $M_1,\dots,M_p$. It was asserted in Cramér [4] that the multiplicity M of the multivariate process $Y(t) = \{Y_1(t),\dots,Y_p(t)\}$ is such that $M \le \sum_{i=1}^{p} M_i$.

This was proved by Ephremides and Thomas [5] under the assumption that the processes $Y_i(t)$, $i = 1,\dots,p$ are pairwise orthogonal. To our knowledge, no proof has appeared in the literature for the general case. In the proof of the next theorem we will use "direct integral" techniques to prove this result for multivariate nonorthogonal random fields. Cramér's assertion then follows as a special case. Although the theorem is stated for generalized fields, there is nothing in the proof that depends on this fact, and the theorem is equally true for ordinary fields.

<u>Theorem 3.3</u>: Let $F_\lambda(\Phi)$, $\lambda \in L$, $\Phi \in K$, be a p.n.d. random field such that L is a separable Hilbert space and $F_\lambda(\Phi)$ is linear in λ for each $\Phi \in K$. Let e_i, $i = 1,2,\dots,$ be an orthonormal basis

in L and denote by $F_i(\Phi)$ the univariate field defined by $F_{e_i}(\Phi)$, $\Phi \in K$. Then $F_i(\Phi)$ is p.n.d., and if d_i and d are the coordinate multiplicities of $F_i(\Phi)$ and $F_\lambda(\Phi)$ respectively, then $d \leqslant \sum_{i=1}^{\dim L} d_i$.

<u>Proof :</u> From our definitions in Section 2 it follows that the coordinate operators of $F_i(\Phi)$ are the coordinate operators of $F_\lambda(\Phi)$ restricted to $H(F_i)$. Let V_i, $i = 1,\dots,m = \dim L$ and V be unitary transformations such that $H(F_i) \overset{V_i}{\longleftrightarrow} \hat{H}_i = \int_{R^n} \hat{H}_i(x)\, d\mu_i(x)$ and $H(F) \overset{U}{\longleftrightarrow} \hat{H} = \int_{R^n} \hat{H}(x)\, d\mu(x)$ are direct integral of Hilbert space decompositions induced by the coordinate operators $\{Q_1,\dots,Q_n\}$ in $H(F_i)$ and $H(F)$ respectively. As in the proof of Theorem 3.1, consider the random variables Z_j^i, $j = 1,\dots,d_i$; $i = 1,\dots,m$ which are realized by the isomorphisms V_i as $Z_j^i \longleftrightarrow \{\hat{Z}_j^i(x)\}$ where $\hat{Z}_j^i(x)$, $j = 1,\dots,\dim \hat{H}_i(x)$ is a basis in $\hat{H}_i(x)$ and $\hat{Z}_j^i(x) = 0$ for $i > \dim \hat{H}_i(x)$. Using Z_j^i we can construct a representation as in (1) satisfying i) and ii) and it is not difficult to prove that $F_i(\Phi)$ is p.n.d. Assume that each $F_i(\Phi)$ is non-zero (otherwise L can be restricted to the subspace spanned by those e_i for $F_i(\Phi)$ is non-zero). Then $d_i \geqslant 1$ for $i = 1,\dots, \dim L$. If L is infinite dimensional or if one of the d_i's is infinite, there is nothing to prove. Assume therefore that each d_i is finite and that L is finite - dimensional with dimension m. Using the same technique as in Theorem 3.2, it follows that each random variable F_i in $H(F_i)$ can be represented as

$$F_i = \sum_{j=1}^{d_i} v_j^i\,(Q_1,\dots,Q_n)\; Z_j^i \tag{4}$$

for some operator functions $v_j^i(Q_1,\dots,Q_n)$. Denote by $\hat{h}_{ij} = \{\hat{h}_{ij}(x)\}$ the realization of Z_j^i by $H(F) \overset{V}{\longleftrightarrow} \hat{H}$. Since $H(F) = \sum_{i=1}^{m} H(F_i)$, where the sum need not be orthogonal, it follows from Eq. (4) that every random variable F in $H(F)$ can be represented as

$$F = \sum_{i=1}^{m} \sum_{j=1}^{d_i} v_j^i (Q_1,\ldots,Q_n) Z_j^i \tag{5}$$

From (4) and (5), $F \overset{V}{\longleftrightarrow} \{\sum_{i=1}^{m} \sum_{j=1}^{d_i} v_j^i(x) \hat{h}_{ij}(x)\}$. Let $\hat{g} = \{\hat{g}(x)\} \in \hat{H}$ be such that $(\hat{g}(x), \hat{h}_{ij}(x))_x = 0$, $i = 1,\ldots,m$; $j = 1,\ldots,\dim\hat{H}_i(x)$. It follows from Eq. (5) that $\hat{g} \perp \hat{F}$ for all $\hat{F} \in \hat{H}$. Thus $\int_{R^n} ||\hat{g}(x)||_x^2 \, d\mu(x) = 0$, and $||\hat{g}(x)||_x^2 = 0$ for μ-measure a.e.
Therefore $\hat{H}(x)$ is spanned by $\hat{h}_{ij}(x)$, $i = 1,\ldots,m$; $j = 1,\ldots, \dim \hat{H}_i(x)$ for μ-measure a.e. This means that

$$\dim \hat{H}(x) \leq \sum_{i=1}^{m} \dim \hat{H}_i(x) \leq \sum_{i=1}^{m} d_i \quad \text{for } \mu\text{-measure a.e.}$$

It follows immediately that $d \leq \sum_{i=1}^{m} d_i$ ||.

Using a similar technique, it can be proved that $\sup_i d_i \leq d$.

4. The Homogeneous Field Case

Let $F_\lambda(\Phi)$, $\lambda \in L$, $\Phi \in K$, be a random field which is homogeneous but not necessarily p.n.d. Let $U_k(t)$, $k = 1,\ldots,n$; $t \in (-\infty,\infty)$, be the unitary strongly continuous groups in $H(F)$ defined by $U_k(t) F_\lambda(\Phi) = F_\lambda(\tau_k(t)\Phi)$ where $\tau_k(t) \Phi(x_1,\ldots,x_k,\ldots,x_n) = \Phi(x_1,\ldots,x_k+t,\ldots,x_n)$. Using Stone's theorem there is a self-adjoint operator P_k such that $U_k(t) = \exp(itP_k)$. The operator P_k will be called <u>kth momentum operator</u> of the field.
One easily verifies that the family $\{P_1,\ldots,P_n\}$ form a commuting family of self-adjoint operators, and this fact was used in [21] to obtain a realization $H(F) \longleftrightarrow \hat{H} = \int_{R^n} H(u) \, d\vartheta(u)$ as a direct integral of Hilbert spaces induced by $\{P_1,\ldots,P_n\}$. Let $A_i = \{u : \dim \hat{H}(u) = i\}$, $i = 0,1,\ldots$ and let A' be the set obtained from $\bigcup_{i=0}^{\infty} A_i$ deleting those A_i with ϑ measure zero.
The <u>momentum multiplicity</u> d_P of the field is defined as $d_P = \sup_{u \in A'} \{\dim \hat{H}(u)\}$. The field is said to have <u>uniform momentum multiplicity</u> if ϑ is equivalent to Lebesgue measure on R^n and $\dim \hat{H}(u) = d_P$ for ϑ-measure a.e. In the case of a random process, $n = 1$, and

following [20] the operator $H = -P$ will be called the energy operator of the process.

For an ordinary homogeneous random field $F_\lambda(x)$, $\lambda \in L$, $x = (x_1,\dots,x_n) \in R^n$, we define the shift operator $U_k(t)$ by $U_k(t)\ F_\lambda(x_1,\dots,x_k,\dots,x_n) = F_\lambda(x_1,\dots,x_k-t,\dots,x_n)$. If $F_\lambda(x)$ is continuous in q.m., Stone's theorem can again be used to define a family of momentum operators $\{P_1,\dots,P_n\}$.
The concept of momentum multiplicity is defined as above.

We turn next to the special case of a p.n.d. and homogeneous random field $F_\lambda(\Phi)$. Then as in the random process case [19] the following commutation relation holds for all $t, s \in (-\infty,\infty)$

$$E_k(t)\ U_k(s) = U_k(s)\ E_k(t+s) \tag{6}$$

where $E_k(t)$ is the projection operator on H_t^k. A relation of type (6) is well known in quantum mechanics and is special case of the imprimitivity relation (Mackey [12, p. 50]). From Theorem 4.1 of [21] it follows that the system of operators $\{P_1,\dots,P_n;Q_1,\dots,Q_n\}$ formed by the momentum-coordinate operators of the field is a so-called direct sum of Schrødinger n-systems[14, p.81]. The following theorem immediately results (see also Corollary 4.1 of [21]).

__Theorem 4.1:__ Let $F_\lambda(\Phi)$, $\lambda \in L$, $\Phi \in K$, be a homogeneous p.n.d. random field with coordinate and momentum multiplicity d_Q and d_P respectively. Let M be the number determined by the decomposition

$$\{P_1,\dots,P_n : Q_1,\dots,Q_n\} = \sum_{j=1}^{M} \oplus \{P_1^j,\dots,P_n^n ; Q_1^j,\dots,Q_n^j\}$$ of the

momentum-coordinate operators as a direct sum of Schrödinger n-systems. Then $d_Q = d_P = M$. Furthermore, the coordinate and momentum multiplicity of $F_\lambda(\Phi)$ are both uniform.
Using the technique of proof in Theorem 4.2 of [21] we have

__Theorem 4.2:__ Let $F_\lambda(\Phi)$, $\lambda \in L$, $\Phi \in K$, be a homogeneous p.n.d. random field with coordinate and momentum multiplicity equal to M. Let Δ, Δ_1 and Δ_2 be Borel sets in R^n of finite Lebesque measure. Then $F_\lambda(\Phi)$ can be represented (* is used to denote the convolution operator)

$$F_\lambda(\Phi) = \sum_{j=1}^{M} \int_{S\_(y(\Phi))} g_\lambda^j(\Phi;\ x)\ dZ_j(x) = \qquad (7)$$

$$\sum_{j=1}^{M} \int_{S\_(y(\Phi))} (G_\lambda^j * \Phi)\ (x) dZ_j(x)$$

where $g_\lambda^j(\Phi;\ \cdot) = (G_\lambda^j * \Phi)(\cdot)$, and G_λ^j, $j = 1,\ldots,M$; $\lambda \in L$, are tempered distributions (that is, continuous linear functionals on K) with support in $S\_(0)$, and where $Z_j(\Delta)$, $j = 1,\ldots,M$ are mutually orthogonal random measures such that

i) $E\{Z_j(\Delta_1)\ \overline{Z_k(\Delta_2)}\} = \delta_{kj}\ |\Delta_1 \cap \Delta_2|$, where $|\cdot|$ denotes Lebesgue measure.

ii) $U(x)\ Z_j(\Delta) = Z_j(\Delta - x)$, where $U(x) = \exp\ (i \sum_{k=1}^{n} x_k P_k)$ for $x = (x_1,\ldots,x_n) \in R^n$.

iii) $H\_(F,y) = \sum_{j=1}^{M} \oplus H\_(Z_j,y)$, where $H\_(Z_j,y)$ is the Hilbert space generated by the linear hull of the set of random variables $Z_j(\Delta)$, when Δ runs through all Borel sets of finite Lebesgue measure such that $\Delta \subset S\_(y)$.

Theorems 4.1 and 4.2 remain true for a ordinary random field $F_\lambda(x)$. For an ordinary field the representation (7) takes the form

$$F_\lambda(y) = \sum_{j=1}^{M} \int_{S\_(y)} g_\lambda^j(y-x)\ dZ_j(x)$$

where the random measures $Z_j(\Delta)$, $j = 1,\ldots,M$ satisfy properties i) - iii) of Theorem 4.2.

Our last result is a relation between the coordinate and momentum multiplicity $M = d_Q = d_P$ and the dimensionality of $F_\lambda(\Phi)$ when $F_\lambda(\Phi)$ satisfies the assumption of Theorem 3.3

<u>Theorem 4.3</u>: Let $F_\lambda(\Phi)$, $\lambda \in L$, $\Phi \in K$, be a homogeneous p.n.d. random field such that L is a separable Hilbert space and $F_\lambda(\Phi)$ is linear in λ for each $\Phi \in K$. Let $m = \dim L$. Then $M = d_Q = d_P \leqslant m$.

Proof: The theorem follows directly from Theorem 3.3 if we are able to prove that a univariate homogeneous p.n.d. field $F(\Phi)$ has $d_Q = d_P = 1$. It follows from Corollary 4.1 of [21] that such a field $F(\Phi)$ has a spectral measure $\tilde{G}$ given by

$$\tilde{G}(\Delta) = \frac{1}{(2\pi)^{2n}} \sum_{j=1}^{M} \int_{\Delta} |\tilde{G}^j(u)|^2 \, du$$

Which is absolutely continuous with respect to Lebesgue measure. Using the technique of the proof of Lemma 2.1 of [2], the absolute continuity of $\tilde{G}$ implies that for an arbitrary fixed $\Phi \in K$, the linear hull of the set of elements $\exp\ (i \sum_{k=1}^{n} x_k P_k)\, F(\Phi)$, where $x = (x_1,\ldots,x_n)$ varies over R^n, is dense in $H(F)$. Thus [16, pp. 383-384] the linear hull of the set of elements $\prod_{k=1}^{n} E^k(\Delta_k)\, F(\Phi)$, where Δ_k, $k = 1,\ldots,n$ vary over the Borel sets of $(-\infty,\infty)$, and E_t^k, $t \in (-\infty,\infty)$, is the resolution of identity associated with P_k, is dense in $H(F)$. It follows that $F(\Phi)$ is a "cyclic element"for the operator system $\{P_1,\ldots,P_n\}$. By a multivariate extension of the arguments in [6, p. 124] we have that $H(F) \leftrightarrow L^2_\Phi(R^n)$ is the space of Borel-measurable functions on R^n which are square integrable with respect to the measure μ_Φ generated by $\mu_\Phi(\Delta) = = E\, |\prod_{k=1}^{n} E^k(\Delta_k)\, F(\Phi)|^2$ for $\Delta = \Delta_1 x \ldots x \Delta_n$. Furthermore, the isomorphism $H(F) \leftrightarrow L^2_\Phi(R^n)$ is a realization of $H(F)$ induced by $\{P_1,\ldots,P_n\}$ according to [13], [21, Theorem 2.1]. Therefore, $d_P = 1$ and the proof is complete.||

Corollary: For a univariate homogeneous p.n.d. field $F(\Phi)$, $\Phi \in K$, we have $M = d_Q = d_P = 1$, and the representation (7) reduces to

$$F(\Phi) = \int_{S_-(y(\Phi))} (G*\Phi)(x)\, dZ(x)$$

This should be compared to the representation obtained in Theorem 7 of Urbanik [22]. Finally we note that Theorem 4.3 remains true for an ordinary random field $F_\lambda(x)$. In this case the proof is simpler. (Compare the proof of Theorem 2 of [19].)
We finally note that by using Mackey's imprimitivity theorem [12, p. 51] in connection with a generalized version of (6) it may be possible to extend some of the results in this section to stochastic processes defined on more general groups than R^n.

Acknowledgement

I am grateful to professor G. Kallianpur for pointing out some mistakes in an earlier version of this paper.

This research was supported by the Advanced Research Projects Agency of the Department of Defense and was monitored by AFTAC, Patrick AFB FL 32925, under Contract No. F08606-77-C-0001.

References

[1] N.I. Achieser and I.M. Glazman, "Theorie der Linearen Operatoren im Hilbert-Raum". Berlin, Akademie-Verlag, 1958.

[2] K. Balagangadharan, " The prediction theory of stationary random distributions". Mem. Coll. Sci., Univ. Kyoto Ser. A, 3(1960), pp. 243-256.

[3] G.Y.H. Chi, "Multiplicity and representation theory of generalized random processes". J. Mult. Anal., 1(1971), pp. 412-432.

[4] H. Cramér, "On the structure of purely non-deterministic processes". Arkiv för Mat., 4 (1961), pp. 249-266.

[5] A. Ephremides and J.B. Thomas, "On the multiplicity of a class of multivariate random processes". Ann. Math. Statist., 42 (1972), pp. 2083-2089.

[6] I.M. Gelfand and N.J. Wilenkin, "Verallgemeinerte Funktionen (Distributionen)" IV. Berlin, Veb Deutcher Verlag der Wissenschaften, 1964.

[7] K. Gustafson and B. Misra, "Canonical commutation relations of quantum mechanics and stochastic regularity". Letters in Mathematical Physics, 1 (1976), pp. 275-280.

[8] O. Hanner, "Deterministic and non-deterministic processes". Ark. Mat., 1 (1950), pp. 169-177.

[9] T. Hida, "Canonical representations of Gaussian processes and their applications". Mem. Coll. Sci., Univ. Kyoto Ser. A, 3 (1960), pp. 109-155.

[10] G. Kallianpur and V. Mandrekar, " Multiplicity and representation theory of purely non-deterministic stochastic processes". Theory Prob.Applications, 10 (1965), pp. 553-581.

[11] P. Masani and J. Robertson, "The time domain analysis of a continuous parameter weakly stationary stochastic process". Pacific J. Math., 12 (1962), pp. 1361-1378.

[12] G.W. Mackey, "Induced representations of groups and quantum mechanics". New York, Benjamin, 1968.

[13] J. von Neumann, "On rings of operators. Reduction theory". Ann. of Math., 50 (1949), pp. 401-485.

[14] C.R. Putnam, "Commutation properties of Hilbert space operators and related topics". New York, Springer, 1967.

[15] M.M. Rao, "Representation theory of multidimensional generalized random fields". In Multivariate Analysis II, P. Krishnaiah, Ed., pp. 411-436.

[16] F. Riesz and B. Sz.-Nagy, "Functional Analysis". New York, Ungar, 1955 (2nd Ed.)

[17] J. Robertson, "Orthogonal decompositions of multivariate weakly stationary stochastic processes". Can J. Math., 20 (1968), pp. 368-382.

[18] D. Tjøstheim, "A note on the unified Dirac-von Neumann formulation of quantum mechanics". J. Math. Phys., 16 (1975), pp. 766-767.

[19] D. Tjøstheim, "A commutation relation for wide sense stationary processes". SIAM J. Appl. Math., 30 (1976), pp. 115-122.

[20] D. Tjøstheim, "Multiplicity theory for multivariate wide sense stationary generalized processes". J. Mult. Anal., 5 (1975), pp. 314-321.

[21] D. Tjøstheim, "Spectral representations and density operators for infinite-dimensional homogeneous random fields". Z. Wahrscheinlichkeitstheorie verw.Gebiete, 36 (1976), pp. 323-336.

[22] K. Urbanik, "A contribution to the theory of generalized stationary random fields". Second Prague Conf.Inf.Theor. Statist., pp. 667-679, New York, Academic Press, 1960.

[23] A.M. Yaglom, "Some classes of random fields in n-dimensional space, related to stationary random processes". Theory Prob. Applications, 2 (1957), pp. 273-320.

[24] K. Yosida, "Functional Analysis". New York, Springer, 1971 (3rd Ed.).

Dag Tjøstheim
NTNF/NORSAR
Post Box 51
N-2007 Kjeller, Norway

Vol. 489: J. Bair and R. Fourneau, Etude Géométrique des Espaces Vectoriels. Une Introduction. VII, 185 pages. 1975.

Vol. 490: The Geometry of Metric and Linear Spaces. Proceedings 1974. Edited by L. M. Kelly. X, 244 pages. 1975.

Vol. 491: K. A. Broughan, Invariants for Real-Generated Uniform Topological and Algebraic Categories. X, 197 pages. 1975.

Vol. 492: Infinitary Logic: In Memoriam Carol Karp. Edited by D. W. Kueker. VI, 206 pages. 1975.

Vol. 493: F. W. Kamber and P. Tondeur, Foliated Bundles and Characteristic Classes. XIII, 208 pages. 1975.

Vol. 494: A Cornea and G. Licea. Order and Potential Resolvent Families of Kernels. IV, 154 pages. 1975.

Vol. 495: A. Kerber, Representations of Permutation Groups II. V, 175 pages. 1975.

Vol. 496: L. H. Hodgkin and V. P. Snaith, Topics in K-Theory. Two Independent Contributions. III, 294 pages. 1975.

Vol. 497: Analyse Harmonique sur les Groupes de Lie. Proceedings 1973–75. Edité par P. Eymard et al. VI, 710 pages. 1975.

Vol. 498: Model Theory and Algebra. A Memorial Tribute to Abraham Robinson. Edited by D. H. Saracino and V. B. Weispfenning. X, 463 pages. 1975.

Vol. 499: Logic Conference, Kiel 1974. Proceedings. Edited by G. H. Müller, A. Oberschelp, and K. Potthoff. V, 651 pages 1975.

Vol. 500: Proof Theory Symposion, Kiel 1974. Proceedings. Edited by J. Diller and G. H. Müller. VIII, 383 pages. 1975.

Vol. 501: Spline Functions, Karlsruhe 1975. Proceedings. Edited by K. Böhmer, G. Meinardus, and W. Schempp. VI, 421 pages. 1976.

Vol. 502: János Galambos, Representations of Real Numbers by Infinite Series. VI, 146 pages. 1976.

Vol. 503: Applications of Methods of Functional Analysis to Problems in Mechanics. Proceedings 1975. Edited by P. Germain and B. Nayroles. XIX, 531 pages. 1976.

Vol. 504: S. Lang and H. F. Trotter, Frobenius Distributions in GL_2-Extensions. III, 274 pages. 1976.

Vol. 505: Advances in Complex Function Theory. Proceedings 1973/74. Edited by W. E. Kirwan and L. Zalcman. VIII, 203 pages. 1976.

Vol. 506: Numerical Analysis, Dundee 1975. Proceedings. Edited by G. A. Watson. X, 201 pages. 1976.

Vol. 507: M. C. Reed, Abstract Non-Linear Wave Equations. VI, 128 pages. 1976.

Vol. 508: E. Seneta, Regularly Varying Functions. V, 112 pages. 1976.

Vol. 509: D. E. Blair, Contact Manifolds in Riemannian Geometry. VI, 146 pages. 1976.

Vol. 510: V. Poènaru, Singularités C^∞ en Présence de Symétrie. V, 174 pages. 1976.

Vol. 511: Séminaire de Probabilités X. Proceedings 1974/75. Edité par P. A. Meyer. VI, 593 pages. 1976.

Vol. 512: Spaces of Analytic Functions, Kristiansand, Norway 1975. Proceedings. Edited by O. B. Bekken, B. K. Øksendal, and A. Stray. VIII, 204 pages. 1976.

Vol. 513: R. B. Warfield, Jr. Nilpotent Groups. VIII, 115 pages. 1976.

Vol. 514: Séminaire Bourbaki vol. 1974/75. Exposés 453 – 470. IV, 276 pages. 1976.

Vol. 515: Bäcklund Transformations. Nashville, Tennessee 1974. Proceedings. Edited by R. M. Miura. VIII, 295 pages. 1976.

Vol. 516: M. L. Silverstein, Boundary Theory for Symmetric Markov Processes. XVI, 314 pages. 1976.

Vol. 517: S. Glasner, Proximal Flows. VIII, 153 pages. 1976.

Vol. 518: Séminaire de Théorie du Potentiel, Proceedings Paris 1972–1974. Edité par F. Hirsch et G. Mokobodzki. VI, 275 pages. 1976.

Vol. 519: J. Schmets, Espaces de Fonctions Continues. XII, 150 pages. 1976.

Vol. 520: R. H. Farrell, Techniques of Multivariate Calculation. X, 337 pages. 1976.

Vol. 521: G. Cherlin, Model Theoretic Algebra – Selected Topics. IV, 234 pages. 1976.

Vol. 522: C. O. Bloom and N. D. Kazarinoff, Short Wave Radiation Problems in Inhomogeneous Media: Asymptotic Solutions. V. 104 pages. 1976.

Vol. 523: S. A. Albeverio and R. J. Høegh-Krohn, Mathematical Theory of Feynman Path Integrals. IV, 139 pages. 1976.

Vol. 524: Séminaire Pierre Lelong (Analyse) Année 1974/75. Edité par P. Lelong. V, 222 pages. 1976.

Vol. 525: Structural Stability, the Theory of Catastrophes, and Applications in the Sciences. Proceedings 1975. Edited by P. Hilton. VI, 408 pages. 1976.

Vol. 526: Probability in Banach Spaces. Proceedings 1975. Edited by A. Beck. VI, 290 pages. 1976.

Vol. 527: M. Denker, Ch. Grillenberger, and K. Sigmund, Ergodic Theory on Compact Spaces. IV, 360 pages. 1976.

Vol. 528: J. E. Humphreys, Ordinary and Modular Representations of Chevalley Groups. III, 127 pages. 1976.

Vol. 529: J. Grandell, Doubly Stochastic Poisson Processes. X, 234 pages. 1976.

Vol. 530: S. S. Gelbart, Weil's Representation and the Spectrum of the Metaplectic Group. VII, 140 pages. 1976.

Vol. 531: Y.-C. Wong, The Topology of Uniform Convergence on Order-Bounded Sets. VI, 163 pages. 1976.

Vol. 532: Théorie Ergodique. Proceedings 1973/1974. Edité par J.-P. Conze and M. S. Keane. VIII, 227 pages. 1976.

Vol. 533: F. R. Cohen, T. J. Lada, and J. P. May, The Homology of Iterated Loop Spaces. IX, 490 pages. 1976.

Vol. 534: C. Preston, Random Fields. V, 200 pages. 1976.

Vol. 535: Singularités d'Applications Differentiables. Plans-sur-Bex. 1975. Edité par O. Burlet et F. Ronga. V, 253 pages. 1976.

Vol. 536: W. M. Schmidt, Equations over Finite Fields. An Elementary Approach. IX, 267 pages. 1976.

Vol. 537: Set Theory and Hierarchy Theory. Bierutowice, Poland 1975. A Memorial Tribute to Andrzej Mostowski. Edited by W. Marek, M. Srebrny and A. Zarach. XIII, 345 pages. 1976.

Vol. 538: G. Fischer, Complex Analytic Geometry. VII, 201 pages. 1976.

Vol. 539: A. Badrikian, J. F. C. Kingman et J. Kuelbs, Ecole d'Eté de Probabilités de Saint Flour V-1975. Edité par P.-L. Hennequin. IX, 314 pages. 1976.

Vol. 540: Categorical Topology, Proceedings 1975. Edited by E. Binz and H. Herrlich. XV, 719 pages. 1976.

Vol. 541: Measure Theory, Oberwolfach 1975. Proceedings. Edited by A. Bellow and D. Kölzow. XIV, 430 pages. 1976.

Vol. 542: D. A. Edwards and H. M. Hastings, Čech and Steenrod Homotopy Theories with Applications to Geometric Topology. VII, 296 pages. 1976.

Vol. 543: Nonlinear Operators and the Calculus of Variations, Bruxelles 1975. Edited by J. P. Gossez, E. J. Lami Dozo, J. Mawhin, and L. Waelbroeck, VII, 237 pages. 1976.

Vol. 544: Robert P. Langlands, On the Functional Equations Satisfied by Eisenstein Series. VII, 337 pages. 1976.

Vol. 545: Noncommutative Ring Theory. Kent State 1975. Edited by J. H. Cozzens and F. L. Sandomierski. V, 212 pages. 1976.

Vol. 546: K. Mahler, Lectures on Transcendental Numbers. Edited and Completed by B. Diviš and W. J. Le Veque. XXI, 254 pages. 1976.

Vol. 547: A. Mukherjea and N. A. Tserpes, Measures on Topological Semigroups: Convolution Products and Random Walks. V, 197 pages. 1976.

Vol. 548: D. A. Hejhal, The Selberg Trace Formula for PSL (2, $\mathbb{R}$). Volume I. VI, 516 pages. 1976.

Vol. 549: Brauer Groups, Evanston 1975. Proceedings. Edited by D. Zelinsky. V, 187 pages. 1976.

Vol. 550: Proceedings of the Third Japan – USSR Symposium on Probability Theory. Edited by G. Maruyama and J. V. Prokhorov. VI, 722 pages. 1976.

Vol. 551: Algebraic K-Theory, Evanston 1976. Proceedings. Edited by M. R. Stein. XI, 409 pages. 1976.

Vol. 552: C. G. Gibson, K. Wirthmüller, A. A. du Plessis and E. J. N. Looijenga. Topological Stability of Smooth Mappings. V, 155 pages. 1976.

Vol. 553: M. Petrich, Categories of Algebraic Systems. Vector and Projective Spaces, Semigroups, Rings and Lattices. VIII, 217 pages. 1976.

Vol. 554: J. D. H. Smith, Mal'cev Varieties. VIII, 158 pages. 1976.

Vol. 555: M. Ishida, The Genus Fields of Algebraic Number Fields. VII, 116 pages. 1976.

Vol. 556: Approximation Theory. Bonn 1976. Proceedings. Edited by R. Schaback and K. Scherer. VII, 466 pages. 1976.

Vol. 557: W. Iberkleid and T. Petrie, Smooth S^1 Manifolds. III, 163 pages. 1976.

Vol. 558: B. Weisfeiler, On Construction and Identification of Graphs. XIV, 237 pages. 1976.

Vol. 559: J.-P. Caubet, Le Mouvement Brownien Relativiste. IX, 212 pages. 1976.

Vol. 560: Combinatorial Mathematics, IV, Proceedings 1975. Edited by L. R. A. Casse and W. D. Wallis. VII, 249 pages. 1976.

Vol. 561: Function Theoretic Methods for Partial Differential Equations. Darmstadt 1976. Proceedings. Edited by V. E. Meister, N. Weck and W. L. Wendland. XVIII, 520 pages. 1976.

Vol. 562: R. W. Goodman, Nilpotent Lie Groups: Structure and Applications to Analysis. X, 210 pages. 1976.

Vol. 563: Séminaire de Théorie du Potentiel. Paris, No. 2. Proceedings 1975–1976. Edited by F. Hirsch and G. Mokobodzki. VI, 292 pages. 1976.

Vol. 564: Ordinary and Partial Differential Equations, Dundee 1976. Proceedings. Edited by W. N. Everitt and B. D. Sleeman. XVIII, 551 pages. 1976.

Vol. 565: Turbulence and Navier Stokes Equations. Proceedings 1975. Edited by R. Temam. IX, 194 pages. 1976.

Vol. 566: Empirical Distributions and Processes. Oberwolfach 1976. Proceedings. Edited by P. Gaenssler and P. Révész. VII, 146 pages. 1976.

Vol. 567: Séminaire Bourbaki vol. 1975/76. Exposés 471–488. IV, 303 pages. 1977.

Vol. 568: R. E. Gaines and J. L. Mawhin, Coincidence Degree, and Nonlinear Differential Equations. V, 262 pages. 1977.

Vol. 569: Cohomologie Etale SGA 4½. Séminaire de Géométrie Algébrique du Bois-Marie. Edité par P. Deligne. V, 312 pages. 1977.

Vol. 570: Differential Geometrical Methods in Mathematical Physics, Bonn 1975. Proceedings. Edited by K. Bleuler and A. Reetz. VIII, 576 pages. 1977.

Vol. 571: Constructive Theory of Functions of Several Variables, Oberwolfach 1976. Proceedings. Edited by W. Schempp and K. Zeller. VI, 290 pages. 1977

Vol. 572: Sparse Matrix Techniques, Copenhagen 1976. Edited by V. A. Barker. V, 184 pages. 1977.

Vol. 573: Group Theory, Canberra 1975. Proceedings. Edited by R. A. Bryce, J. Cossey and M. F. Newman. VII, 146 pages. 1977.

Vol. 574: J. Moldestad, Computations in Higher Types. IV, 203 pages. 1977.

Vol. 575: K-Theory and Operator Algebras, Athens, Georgia 1975. Edited by B. B. Morrel and I. M. Singer. VI, 191 pages. 1977.

Vol. 576: V. S. Varadarajan, Harmonic Analysis on Real Reductive Groups. VI, 521 pages. 1977.

Vol. 577: J. P. May, E_∞ Ring Spaces and E_∞ Ring Spectra. IV, 268 pages. 1977.

Vol. 578: Séminaire Pierre Lelong (Analyse) Année 1975/76. Edité par P. Lelong. VI, 327 pages. 1977.

Vol. 579: Combinatoire et Représentation du Groupe Symétrique, Strasbourg 1976. Proceedings 1976. Edité par D. Foata. IV, 339 pages. 1977.

Vol. 580: C. Castaing and M. Valadier, Convex Analysis and Measurable Multifunctions. VIII, 278 pages. 1977.

Vol. 581: Séminaire de Probabilités XI, Université de Strasbourg. Proceedings 1975/1976. Edité par C. Dellacherie, P. A. Meyer et M. Weil. VI, 574 pages. 1977.

Vol. 582: J. M. G. Fell, Induced Representations and Banach *-Algebraic Bundles. IV, 349 pages. 1977.

Vol. 583: W. Hirsch, C. C. Pugh and M. Shub, Invariant Manifolds. IV, 149 pages. 1977.

Vol. 584: C. Brezinski, Accélération de la Convergence en Analyse Numérique. IV, 313 pages. 1977.

Vol. 585: T. A. Springer, Invariant Theory. VI, 112 pages. 1977.

Vol. 586: Séminaire d'Algèbre Paul Dubreil, Paris 1975–1976 (29ème Année). Edited by M. P. Malliavin. VI, 188 pages. 1977.

Vol. 587: Non-Commutative Harmonic Analysis. Proceedings 1976. Edited by J. Carmona and M. Vergne. IV, 240 pages. 1977.

Vol. 588: P. Molino, Théorie des G-Structures: Le Problème d'Equivalence. VI, 163 pages. 1977.

Vol. 589: Cohomologie l-adique et Fonctions L. Séminaire de Géométrie Algébrique du Bois-Marie 1965–66, SGA 5. Edité par L. Illusie. XII, 484 pages. 1977.

Vol. 590: H. Matsumoto, Analyse Harmonique dans les Systèmes de Tits Bornologiques de Type Affine. IV, 219 pages. 1977.

Vol. 591: G. A. Anderson, Surgery with Coefficients. VIII, 157 pages. 1977.

Vol. 592: D. Voigt, Induzierte Darstellungen in der Theorie der endlichen, algebraischen Gruppen. V, 413 Seiten. 1977.

Vol. 593: K. Barbey and H. König, Abstract Analytic Function Theory and Hardy Algebras. VIII, 260 pages. 1977.

Vol. 594: Singular Perturbations and Boundary Layer Theory, Lyon 1976. Edited by C. M. Brauner, B. Gay, and J. Mathieu. VIII, 539 pages. 1977.

Vol. 595: W. Hazod, Stetige Faltungshalbgruppen von Wahrscheinlichkeitsmaßen und erzeugende Distributionen. XIII, 157 Seiten. 1977.

Vol. 596: K. Deimling, Ordinary Differential Equations in Banach Spaces. VI, 137 pages. 1977.

Vol. 597: Geometry and Topology, Rio de Janeiro, July 1976. Proceedings. Edited by J. Palis and M. do Carmo. VI, 866 pages. 1977.

Vol. 598: J. Hoffmann-Jørgensen, T. M. Liggett et J. Neveu, Ecole d'Eté de Probabilités de Saint-Flour VI – 1976. Edité par P.-L. Hennequin. XII, 447 pages. 1977.

Vol. 599: Complex Analysis, Kentucky 1976. Proceedings. Edited by J. D. Buckholtz and T. J. Suffridge. X, 159 pages. 1977.

Vol. 600: W. Stoll, Value Distribution on Parabolic Spaces. VIII, 216 pages. 1977.

Vol. 601: Modular Functions of one Variable V, Bonn 1976. Proceedings. Edited by J.-P. Serre and D. B. Zagier. VI, 294 pages. 1977.

Vol. 602: J. P. Brezin, Harmonic Analysis on Compact Solvmanifolds. VIII, 179 pages. 1977.

Vol. 603: B. Moishezon, Complex Surfaces and Connected Sums of Complex Projective Planes. IV, 234 pages. 1977.

Vol. 604: Banach Spaces of Analytic Functions, Kent, Ohio 1976. Proceedings. Edited by J. Baker, C. Cleaver and Joseph Diestel. VI, 141 pages. 1977.

Vol. 605: Sario et al., Classification Theory of Riemannian Manifolds. XX, 498 pages. 1977.

Vol. 606: Mathematical Aspects of Finite Element Methods. Proceedings 1975. Edited by I. Galligani and E. Magenes. VI, 362 pages. 1977.

Vol. 607: M. Métivier, Reelle und Vektorwertige Quasimartingale und die Theorie der Stochastischen Integration. X, 310 Seiten. 1977.

Vol. 608: Bigard et al., Groupes et Anneaux Réticulés. XIV, 334 pages. 1977.